PRESENT
at the
CREATION

Also by Amir D. Aczel:

Fermat's Last Theorem

Probability 1

God's Equation

The Mystery of the Aleph

The Riddle of the Compass

Entanglement

Pendulum

Chance

Descartes' Secret Notebook

The Artist and the Mathematician

The Jesuit and the Skull

The Cave and the Cathedral

Uranium Wars

PRESENT
at the
CREATION

The Story of CERN
and the Large Hadron Collider

Amir D. Aczel

Crown Publishers
New York

Copyright © 2010 by Amir D. Aczel

All rights reserved.
Published in the United States by Crown Publishers, an
imprint of the Crown Publishing Group, a division of
Random House, Inc., New York.
www.crownpublishing.com

CROWN and the Crown colophon are registered trademarks
of Random House, Inc.

Grateful acknowledgment is made to Ruth Braunizer for
permission to reprint a poem by Erwin Schrödinger written
in 1942, which now serves as his epitaph. Translated into
English by Arnulf Braunizer.

Library of Congress Cataloging-in-Publication Data
Aczel, Amir D.
  Present at the creation : the story of CERN and the large
hadron collider / by Amir Aczel. — 1st ed.
    p.   cm.
  Includes bibliographical references.
  1. Large Hadron Collider (France and Switzerland)
2. Colliders (Nuclear physics) 3. European Organization for
Nuclear Research. I. Title.
  QC787.P73A29 2010
  539.7'36094—dc22   2010014835

ISBN 978-0-307-59167-8

Printed in the United States of America

Book design by Donna Sinisgalli

10 9 8 7 6 5 4 3 2 1

First Edition

For Miriam,

who loves physics

and marvels at the mysteries

of the universe

Contents

Acknowledgments

Only once or twice in a lifetime may a writer encounter an opportunity with the potential to change worldwide understanding of science and in this way touch the lives of many people. The fantastic story of the Large Hadron Collider (LHC) at the international laboratory called CERN, near Geneva, Switzerland, and how it is changing the way we view the universe is such an opportunity for me, one that has gripped me and taken me on a passionate voyage of discovery. But the project of researching and telling this story would not have been possible without the help, care, encouragement, and enthusiasm of many individuals from around the world.

Researching this odyssey of science has been immensely rewarding for me, and I am very grateful to the many leading scientists who have been so generous with their time and effort in explaining to me in great detail their important contributions to knowledge as well as their ongoing work in science and their aspirations for the future.

These people include thirteen Nobel Prize winners in physics, and a score of other leading physicists, cosmologists, and mathematicians. They are, collectively, the crème de la crème of twenty-first-century physical science. I am grateful to all of them for their support, generosity, and interest in this undertaking.

I've been very fortunate to be invited several times to visit CERN, and I thank my friend the mathematician, physicist, and intellectual Carlo F. Barenghi of Newcastle University for arranging my invitation to the laboratory. At CERN, my deep gratitude goes to the Italian physicist Paolo Petagna of the Compact Muon Solenoid (CMS) group for showing me around, introducing me to many scientists, and enabling me to study the minutest details of many of the components of the collider and its detectors, as well as sharing with me much of the history and the magic of CERN.

I thank Jack Steinberger, Nobel Laureate in physics, for telling me about his codiscovery of the muon neutrino, and for explaining to me the early history of the CERN laboratory. Also at CERN, I am very grateful to Fabiola Gianotti, spokesperson of ATLAS, for a discussion of the goals of the ATLAS collaboration and her life in physics. I am indebted to Peter Jenni, former ATLAS spokesperson, for information on the work of ATLAS. Also at ATLAS, my warm thanks go to Manuela Cirilli for showing me the ATLAS control room and telling me about her career in physics.

It is my great pleasure to acknowledge my debt to Guido Tonelli, the spokesperson of the CMS group at CERN, for making me feel so at home at his center, for explaining to me how the CMS detector works, and for sharing with me the dramatic results of particle collisions at 7 TeV as soon as they were obtained on March 30, 2010. I am very grateful to Stefano Redaelli for showing me how the LHC is controlled and telling me about his work at the CERN Control Center. I am grateful to Paola Tropea for showing me around the CERN Control Center, and to Peter Sollander for explaining to me the CERN cryogenics and the control of temperatures in the accelerators, as well as many aspects of the infrastructure of the great machine. I thank Jasper Kirkby for explaining to me the CLOUD project at CERN, and Roberto Corsini for showing me the prototype for CERN's next linear accelerator. I am grateful to Luis Alvarez-Gaume, head of the CERN Theory Division, for his time and for sharing with me his work and ideas, and I am thankful to the theoretical physicist John Ellis for suggestions and information about his work at CERN and for sketching for me a number of his famous penguin diagrams. My gratitude goes to the writer Rebecca Sara Leam and to communications director James Gillies of the CERN Communication Group for their interest in my work and for welcoming me to the center. I thank CERN for permission to reproduce a number of stunning images.

I am grateful to Leon Lederman, Nobel Laureate in physics, of Fermilab and the University of Chicago, codiscoverer of the muon neutrino, for telling me the details of his great achievements and sharing his scientific explorations.

I am greatly indebted to Steven Weinberg, Nobel Laureate in physics, for inviting me to visit him at the University of Texas at Austin and for explaining the fascinating story of his Nobel work on unifying the electromagnetic and the weak interactions—one of the greatest achievements in theoretical physics. My many thanks go to Gerard 't Hooft, Nobel Laureate in physics, of Utrecht University, for enlightening me about his work on renormalizing gauge theories and his present work on the fundamental elements of quantum theory. I thank Sheldon L. Glashow, Nobel Laureate in physics, of Boston University, for sharing his work in theoretical physics.

At Stanford, California, I am grateful to Martin Perl, Nobel Laureate in physics, of the Stanford Linear Accelerator Center (SLAC), for warmly welcoming me to the laboratory and for recounting the wonderful story of his discovery of the tau lepton. Also at Stanford, my gratitude goes to Leonard Susskind, theoretical physicist extraordinaire, for sharing his discoveries about black holes, strings, the holographic principle he codeveloped with Gerard 't Hooft, and other topics of modern physics.

At Berkeley, my gratitude goes to George Smoot, who won the 2006 Nobel Prize in Physics for discovering the ripples in the microwave background radiation that permeates space, for explaining his findings, and to Mary K. Gaillard for discussing with me her results on predicting the mass of the charm and bottom quarks.

A number of physicists at MIT have been especially helpful to me in preparing this book. I am indebted to Frank Wilczek, Nobel Laureate in physics, for sharing with me the tale of his codiscovery with David Gross and Hugh David Politzer of the mechanism of asymptotic freedom, which governs the behavior of quarks and their eternal confinement inside hadrons. With great pleasure I warmly thank my friend Jerome I. Friedman, Nobel Laureate in physics, for describing to me his key work at the SLAC accelerator on the first experimental proof of the existence of quarks, for countless helpful suggestions and clarifications, and for reading the entire manuscript. I also thank Daniel Freedman for a short discussion of symmetry in particle physics. My deep gratitude goes to Alan Guth, the discoverer of cosmic inflation, for explaining many aspects of gauge theories. I am very grateful to Wolfgang Ketterle, Nobel Laureate in physics, for sharing with me details of his achievement of the first Bose-Einstein condensate and work on fermion gases and for many ideas and insights. It is my great pleasure to acknowledge the help of my friend Barton Zwiebach, string theory expert, in explaining to me many elements of particle theory, strings, and the physics of the Large Hadron Collider, and I thank him warmly for going over the manuscript of the book, pointing out where ideas could be improved and made more precise.

At Princeton, I am greatly indebted to Philip W. Anderson, Nobel Laureate in physics, for a discussion of condensed-matter physics and its relation to the so-called Higgs mechanism and for explaining his work in this area. I am indebted to the late John Archibald Wheeler for explaining to me the intricacies of quantum magic during a visit to his summer home in Maine.

At the Hebrew University of Jerusalem, I am most grateful to Jacob Bekenstein for a deep discussion of his work on black holes and their most bizarre properties.

In Paris, I thank quantum theorist Alain Aspect and Nobel Laureate Claude Cohen-Tannoudji for comprehensive discussions of quantum mechanics. My many thanks go to Gabriele Veneziano of the Collège de France, the inventor of string theory, for discussing with me his theory as well as many aspects of the Standard Model of particle physics and the workings of the Large Hadron Collider. In Oxford, England, I am indebted to the mathematician Sir Roger Penrose for sharing his ideas about black holes and string theory. In Vienna, I am grateful to my friend Anton Zeilinger for discussions of the quantum properties of particles; and in Berlin I am indebted to my friend Jürgen Renn for many facts about the life of Einstein. In Washington, DC, I am thankful to William Phillips, Nobel Laureate in physics, of NIST, for a discussion of the quantum properties of laser-cooled atoms.

I am indebted to the American Institute of Physics (AIP) and the Niels Bohr Library and Archives of AIP's Center for History of Physics; to Gregory Good, director of the Center for History of Physics; and to the center's librarians and archivists, especially Melanie Brown, Scott Prouty, Jennifer Sullivan, and Julie Gass, for research help during my visit to the Niels Bohr Library in June 2009. I also thank the Friends of the Center for History of Physics for a grant-in-aid that allowed me to travel to the Niels Bohr Library and to benefit from perusing its rich primary sources on the history of physics.

I am indebted to the organizers, attendees, and presenters at a number of major professional physics conferences over the years, especially those of the Harvard-MIT 2009 conference "Perspectives in Mathematics and Physics" and Northeastern University's SUSY09 and BSM-LHC conferences in 2009, for the opportunity to learn much about leading-edge research related to the LHC, supersymmetry, string theory, and physics beyond the Standard Model.

I am grateful to Boston University physicist Jabeen Shabnam, working at Fermilab in Illinois, for helping me understand some notions of particle physics. Equally, I am grateful to physicists David Straub, Michael Wick, and Wolfgang Altmannshofer, of the Technische Universität München, for a lively and enlightening discussion of supersymmetry and related models. I thank Haleh Hadavand of Southern Methodist University for providing me with information about particle research work at the ATLAS and CMS detectors. I am indebted to Cumrun Vafa of Harvard University for sharing with me a presentation on string theory. I thank Marie Musy of Dijon for her views on the LHC and its safety.

In Alpbach, in the Austrian Alps, I am most grateful to Ruth and Arnulf Braunizer, Erwin Schrödinger's daughter and son-in-law, for welcoming me to their house, Das Schrödinger Haus, and showing me many of the documents left behind by the great physicist. And in the Swiss Alps, my heartfelt thanks go to my friend Alfonso de Orleans-Borbon for hosting me in his house while on my visits to CERN.

This book would not have been possible without the encouragement, support, and enthusiasm of my friend and agent, Albert Zuckerman of Writers House, and I am immensely thankful to him for his crucial help to me throughout the entire process of turning an idea into a book and for his active involvement in the book's conception, development, and production. I thank Maya Rock of Writers House for her help.

My very deep and warm gratitude goes to Peter Guzzardi, the best editor of all time, for his incredibly insightful, careful, and thoughtful editing of the manuscript in all its stages, and for his countless suggestions and ideas that shaped this book into what it is. My sincere thanks go to my editor at the Crown Publishing Group, John Glusman, for seeing this complicated project through to its successful completion and for working hard to promote this book. Also at Crown, I am most grateful to Domenica Alioto for all her work in helping develop and produce the book, as well as the rest of the production team, including Mary Anne Stewart, Rachelle Mandik, Songhee Kim, and Norman Watkins. My deep gratitude and affection go to my wife, Debra, and my daughter, Miriam, for their immensely helpful suggestions and ideas, and I thank Miriam for research assistance.

The characters in this book are without exception among the most important physicists, cosmologists, and mathematicians the world has ever known, and it has been a great honor and an enviable privilege for me to get to know many of them personally and to learn about the work of others. From the first great physicist I met, Werner Heisenberg, in 1972, to the last one I interviewed, Gerard 't Hooft, in 2010, interacting with these giants of science, ever on a search to uncover the deepest mysteries of the universe, has been an incredible, awe-inspiring experience. My hope for this book is that the reader will share in the excitement of the quest.

Chapter 1

The Exploding Protons

During a number of milestone events in the recent history of our planet, Stefano Redaelli, a tall, thin, bearded thirty-three-year-old particle physicist from Milan with keen eyes and an easy smile, has been at the controls. Some would even say that on these occasions, when the gargantuan particle accelerator known as the Large Hadron Collider (LHC) is being powered to energy levels so immense they have never been seen before, Redaelli is not only the most powerful man who ever lived, but also the only person in history who, with a click of a mouse, could alter forever the fate of the world, and perhaps even of the entire solar system.

At 4:40 p.m. on Friday, March 5, 2010, Redaelli was once again the engineer in charge at the CERN Control Center outside the French village of Prévessin, just across the Swiss border from the headquarters of CERN, the European Organization for Nuclear Research. This is the place that governs the operation of the Large Hadron Collider, the most powerful particle accelerator in the world, as well as the series of smaller accelerators successively feeding the LHC with faster and faster protons (positively charged particles). It is from here that the LHC had just restarted after its winter break, incrementally increasing its power to new records.

This time the world's news media had been kept away from the collider as it powered up, but by a stroke of luck I was allowed access to this nerve center of the entire LHC operation. I looked around me. I was in an ultramodern space about the size of a basketball court, one of whose walls had windows that reached all the way up to the ceiling, framing the snow-capped mountains of the French Jura in the near distance. Arrayed along the other walls were dozens of large, colorful display screens. Scientists and engineers clustered around four large knots of tables laden with computer consoles. The control center looked like a cross between the flight deck of the starship *Enterprise* and the floor of the New York Stock Exchange, but the big screens along the walls, on which Redaelli and his colleagues were now focused intently, were not displaying readouts from deep space or the latest stock prices. Instead they registered a stream of precise data that originated deep inside a circular tunnel measuring sixteen and a half miles in length, buried 300 feet below us. These measurements included: temperature—the lowest in the known universe, colder than the temperature of outer space; magnetic field strengths—among the most powerful ever created by man, some of them more than 200,000 times that of Earth's magnetic field; and energy—at this moment 450 gigaelectron volts (GeV), an extraordinary level that would eventually ramp up to the almost inconceivable 7 teraelectron volts (TeV), which is more than fifteen times as high.[1]

As the engineer in charge, Redaelli was the man whose commands produced the energy increases inside the tunnel below us by raising electric power, now within the green range on one of the large screens, to yellow (and in unusual circumstances to red) at hundreds of megawatts—the power consumption of a medium-size city. The electric current, fed into some 10,000 giant superconducting magnets and radio frequency devices, concentrates, bends, and accelerates the LHC's twin proton beams, eventually raising their speeds to levels extremely close to that of light.

There were many other young scientists in the room, including Peter Sollander, a tall, bespectacled young technical expert from Sweden who was in charge of part of the infrastructure of the collider.

Next to his area was the center controlling the liquid helium cooling the superconducting magnets in the tunnel. Each bar on a screen on the wall before us represented 154 magnets, and all the bars were now green, indicating that none of the temperature measurements from the magnets underground exceeded 1.9 degrees Kelvin (that is, 1.9 degrees Celsius above absolute zero, or –456.25 degrees Fahrenheit). This is the ambient temperature for superconducting magnets. Should the temperature in any magnet rise above its present level, its bar would turn red, and the entire operation would immediately have to shut down to prevent a disaster.

Other scientists were monitoring various aspects of the control of the most complicated scientific operation ever undertaken. On the left side of this large room was a subcenter for the feeder accelerators, which contributed power in stages. The first was a linear beam accelerator called Linac2, and it was followed by the more powerful Proton Synchrotron Booster, then by the Proton Synchrotron itself, and finally by the Super Proton Synchrotron (SPS)—a machine with a celebrated history of discoveries in particle physics in the 1980s. This last accelerator fed protons directly into the Large Hadron Collider. Another cluster of consoles controlled all technical aspects of the giant magnets underground and the electric power flowing into them. The last cluster on the right, where Redaelli was standing, was the control center for the LHC itself.

Right behind the young scientists huddled around the computer screens in this part of the room stood a stern-faced man in his sixties with wavy gray hair, wearing a light blue sweater and jeans, his eyes fixed on the third screen from the left on the wall above. Lyndon ("Lyn") Evans was the silent power, the éminence grise of the control room. He was watching a blue line on the screen, which represented the power driving two opposing beams of protons racing around the 16.5-mile circuit underground at near light speed. Evans, a Welsh physicist known at CERN as "the father of the LHC," represented the organization's top management, but as is typical in this highly unusual international collaboration of more than ten thousand scientists from around the world, the actual decisions were often left to the young people here: the scientists and engineers who run the day-to-day operation of the collider.

At the same time that Redaelli and his colleagues were controlling the Large Hadron Collider from the CERN Control Center, still other scientists were manning the collider's four ultramodern control hubs that govern the actual scientific experiments being carried out in the LHC. One of these state-of-the-art control rooms was located about five miles to the west, at "Point 5" of the LHC, right above a giant detector called CMS (for Compact Muon Solenoid). Here Dr. Guido Tonelli, a leading particle physicist from Pisa, was controlling the action as his group of scientists watched their screens and waited to hear from the CERN Control Center at Prévessin whether the protons accelerated in the tunnel would be allowed to crash at high energy in the superconducting detector right below their feet. Tonelli was scrutinizing information on a computer screen as if oblivious to the rest of the room—crammed with other monitors, cables, and sophisticated computer equipment.

The heaviest scientific instrument ever built, the CMS is a gigantic construct of steel, copper, gold, silicon, many thousands of lead-tungstate crystals, and miles of superconducting niobium-titanium coils, as well as a reservoir of liquid helium; it is densely packed with extremely sensitive complex electronics, and it weighs a total of 12,500 tons. Just the iron inside the CMS detector weighs 10,000 tons—more than the weight of the Eiffel Tower. The outer shell of the huge device is a very powerful magnet, a superconducting electromagnet that must be cooled by liquid helium to a temperature below that of outer space in order to maintain its superconductivity—the conduction of electricity without resistance—required to power the magnet to the very high level of 4 tesla (a hundred thousand times Earth's magnetic field; some magnets performing other tasks in the LHC produce a magnetic field strength twice as high). The energies of the particles that explode inside the CMS detector have not been seen since a trillionth of a second after the Big Bang launched our universe 13.7 billion years ago.

I had gone to the CMS control center an hour earlier this day, and standing inside this room, I couldn't help but wonder at the incongruity of it all. This control room was housed in a building standing all alone in the middle of the bucolic French countryside, surrounded by cow pastures and plowed tracts of land, half a mile from the small village of Cessy. The nearest town was four miles to the southeast: Ferney-Voltaire. (The name Voltaire had been added to Ferney to commemorate the fact that in the eighteenth century, the famous French writer and philosopher lived here, wrote *Candide*, and contributed greatly to the economy of the town.)

Just outside Ferney-Voltaire was "Point 8" of the LHC, the location of a special-purpose detector called LHCb ("b" stands for "beauty"). Farther to the southeast was the Swiss border, and beyond it the suburbs of Geneva. At Meyrin, a western suburb near the Geneva Airport, was "Point 1" of the LHC, the location of a detector named ATLAS (A Toroidal LHC ApparatuS), whose function was similar to that of CMS and whose team of scientists was pursuing similar experiments with crashing protons; and near it was the sprawling headquar-

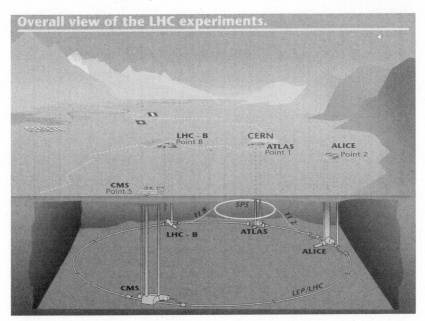

Overall view of the LHC experiments.

ters of CERN. If one continued west along the circular track of the LHC, again crossing into France, within a few miles one would reach "Point 2," the location of the last main detector of the LHC, called ALICE (A Large Ion Collider Experiment), which, like LHCb, was designed for a special scientific purpose.

Some months earlier, on November 30, 2009, just before the operation of the LHC was stopped for a winter break, Tonelli and his team of young scientists had been following tracks on their screens that represented the passage of thousands of tiny particles cascading from the first head-on collisions of protons traveling toward each other at nearly the speed of light and then exploding with immense energies inside the giant underground Compact Muon Solenoid detector.

When the CMS detector is operating, billions of protons crash inside it every second. Tonelli explained to me that of these, only one in a hundred thousand represents an "unusual event" that could potentially be of great interest to science. Higher-level algorithms are used to further skim the sample, and only 300 events per second are permanently recorded to undergo complete reconstruction and physics analysis. Out of these, about 1 high-interest particle collision is shown on screen every second, since our eyes can't perceive very well a complicated image that lasts less time than that.[2] Scientists operate the control room around the clock, and every once in a while they see a spectacular cascade of particles—debris from the immense explosion of protons crashing head-on in the belly of the giant detector deep underground below them. In November and December 2009, before the machine was shut down, there were many such trails of particles from proton collisions at record high energies.

What did they represent? Could they have been the signature of the elusive Higgs boson, a particle that physicists believe is responsible for endowing all matter in the universe with its mass, the so-called God Particle? Was there a hint of the existence of the unseen and mysterious "dark matter" that physicists and astronomers believe permeates galaxies? Or did the detector record the telltale sign of a hidden dimension of the space we live in, one of six or seven additional dimensions suggested by string theory? Any one of these discoveries would represent a giant step forward in our understanding of nature, and all of them are among the goals that the Large Hadron Collider was built to achieve.

The intense amount of highly concentrated energy released by proton collisions inside the LHC takes science to an unexplored new level, a region of high energy the like of which has not been seen in our universe since a fraction of a second after the Big Bang. In this way the Large Hadron Collider is taking us back billions of years to conditions that prevailed in the universe shortly after its fiery birth. Thanks to the LHC, physical science will never be the same as we peer far deeper into the universe than ever before; uncover its structure, past and present; glimpse its future; and perhaps even decipher its meaning.

The head-on collisions of trillions of protons taking place deep under the ground of the border region of Switzerland and France turn energy into mass in the form of other particles that emanate from the collisions and fly off in various directions at high speeds. This process occurs because of Einstein's famous equation, $E = mc^2$, which says that mass and energy are merely two different manifestations of the same thing. Einstein's incredibly powerful formula (actually, a variation that is somewhat more complicated and incorporates the speed of the particles) is what makes all research in particle accelerators possible.[3] The idea is as follows.

Particles are accelerated to great speeds and then made to crash into incoming particles from the opposite direction. Energy is released from these collisions, and in accordance with Einstein's formula, this energy then turns into other fast-moving particles.[4] So from the released energy that resulted from the particle collisions, mass can be *created*. This new mass, born of pure energy, may constitute particles like those that existed when the universe was only a tiny fraction of a second old, and studying their behavior holds the key to our understanding of the forces and particles we see in the world today.

The LHC thus re-creates particles and natural phenomena that have never before been observed. It also takes us back in time to a very distant primordial past when the universe was an immensely dense and hot "soup" of particles, called the *quark-gluon plasma*. The collider also acts as a giant microscope: It can show us the inner workings of space-time.

The Large Hadron Collider was designed and constructed over a period of twenty years at immense expense—costs have by now exceeded $10 billion—by CERN scientists with one aim in mind: to uncover the ultimate laws of the universe. To discover these laws and see long-lost particles, forces, and interactions required an unprecedented effort that could only be undertaken through close international cooperation spanning many institutions and countries and areas of scientific expertise. The LHC project is the most advanced scientific cooperation in history.

How does the machine work? In the LHC experiment, two beams of protons obtained by ionizing hydrogen gas are made to travel in opposite directions with continuously increasing speed. The machine is called the Large *Hadron* Collider because protons are hadrons. A *hadron* (from the Greek word for "thick") is a particle made of quarks. A proton is made up of three quarks, and it therefore belongs to a more specific category of hadrons, called the *baryons*. Hadrons that contain only two quarks are called *mesons*. The protons produced from the hydrogen gas are gradually accelerated in a successive series of CERN's smaller accelerators, until they reach a speed at which they can be injected into the LHC. Here, powerful radio frequency devices in the tunnel "kick" the particles faster every time they pass by. Giant superconducting electromagnets, cooled to almost absolute zero in order to give them electrical conductivity with zero resistance and thereby endow them with the maximum possible power, bend the paths of the protons along the circular trajectory underground and concentrate and maintain the two opposing beams.

The LHC uses 9,593 superconducting electromagnets: There are 1,232 main magnets curving the trajectory of the protons along the underground racecourse, 392 magnets focusing the beams of protons around the tunnel, and 6,400 corrector magnets, making small adjustments in the paths of the protons so that they will crash at locations determined with a precision of a small fraction of a millimeter—much less than the width of a human hair. Still other magnets, some of them embedded inside the main magnets, perform related tasks.

When operated at its maximum energy level, the LHC keeps accelerating the protons until they reach the almost unimaginable speed of 99.9999991 percent of the speed of light (which is 186,282.397 *miles per second*). This happens when the LHC is run at an energy level of 14 TeV (teraelectron volts). One TeV is roughly the energy of the flight of a mosquito, which would seem like a tiny amount—but it is highly concentrated: The LHC generates fourteen times this energy in the volume of *a pair of protons,* which means it is packed into a space a *trillion times* smaller than a mosquito.[5] This is by far the highest energy level per volume ever achieved.[6] In this extremely high-energy realm, new particles and phenomena are likely to appear that until now have lived only in physicists' imaginations.

The Nature of the Quest

For some time now, our progress in trying to understand nature in its most fundamental form has been stymied. We have reached a plateau in our quest for answers to the deepest mysteries of nature because our laboratories—particle accelerators and other means of studying small particles—have yielded little that is new. In particular, the last particle needed to complete and confirm the validity of the *Standard Model of particle physics,* the highly trusted twentieth-century theory that has led to many accurately verified predictions, has yet to be found. This missing ingredient of the Standard Model is called the Higgs boson, also known as the God Particle, because it is believed to imbue all particles with their mass. Physicists believe they have come close to finding it, and there are strong experimental hints that it actually exists, but so far the Higgs has eluded all our intense efforts to discover it.[7]

The Standard Model consists of "matter" particles called fermi- ons (in honor of Enrico Fermi): the electron; the quarks that make up protons and neutrons—the components of atomic nuclei; and similar elementary particles. Then there are "force-carrying" particles, called bosons (in honor of Satyendra Nath Bose), which carry out the work of the forces of nature: gravity, electromagnetism, the weak nuclear force, and the strong nuclear force. The Standard Model accounts for the ac- tions of three of the four forces that act on matter particles through the bosons; gravity is not yet part of the model.

The Standard Model, which has been developed throughout the twentieth century, is one of the most successful scientific theories in his- tory. It is based on quantum theory, the concept of fields, and Einstein's *special* theory of relativity—three pillars of modern physics. However, physicists have not yet found a way to incorporate the fourth pillar, Einstein's *general* theory of relativity—the modern theory of gravity that extends Newton's mechanics to the realms of extremes of speed and gravitational intensity—into the Standard Model, and achieving this goal is one of science's greatest dreams.

Although the success of the Standard Model in explaining the be- havior of particles and forces is awe inspiring, the theory itself lacks one last piece of verification—an experimental finding of the Higgs boson. The Higgs, which the LHC searches for, is the result of theo- retical research in the 1960s, which led to an understanding that there is a mechanism, called the Higgs mechanism, by which the massive particles obtained their masses. This mechanism is believed to have worked its magic in the very early moments after the Big Bang through the mediation of an invisible particle now known as the Higgs boson. The LHC searches for the Higgs boson by re-creating the conditions of high density and immensely high temperatures that prevailed in the early universe. If the Higgs is found, the discovery would provide the final proof of the validity of the Standard Model and would constitute an enormously important result for science.

That being said, we also know from various indications—such as the fact that the forces of nature should unite under very high energy, meaning at a period of time that followed very closely after the Big Bang, but under the Standard Model they do not—that there is place for physics beyond the Standard Model. Modern particle physics, including the Standard Model, is built on the powerful mathematical ideas of symmetry, as we will see in detail. But the hoped-for new physics may well rely on a larger theory that extends the ideas of symmetry to something called *supersymmetry*—a powerful larger symmetry that encompasses both matter particles (the fermions) and the force-carrying particles (the bosons). Thus far, only indirect hints about particle behavior linked to supersymmetry have been glimpsed, and particles that fall under this mathematically advanced view of nature remain to be uncovered.

The discovery of a single supersymmetric particle would be as groundbreaking for science as finding the Higgs; some say much more so. The Large Hadron Collider is believed capable of creating conditions that will allow such theoretical particles to appear. And the discovery of such a "superpartner" particle holds promise of explaining one of the most persistent mysteries in physics and astronomy: the existence of the spooky "dark matter," first theorized in the 1930s, believed to permeate all the galaxies in the universe.

Equally, there is a possibility that nature has hidden dimensions above the usual three of space and one of time, and these other dimensions may also be awaiting discovery. String theory, a view of physics that considers elementary particles as tiny vibrating strings, posits that nature must have as many as ten or eleven dimensions. String theorists hope that the LHC work will lead to the discovery of at least one additional dimension of space-time. This, too, would constitute a major advance in science, implying that the structure of the universe is far more complex than indicated by our present understanding.

We have many great remaining questions about nature that beg to be answered. At its most fundamental level, what is matter made of? What gives matter its mass? What is the nature of the forces that work on matter, and how do they relate to one another? Why are some forces weak and others strong? Are the forces mere manifestations of one unified larger force, called the *superforce,* from which they evolved? What is the dark matter that permeates the universe? What is the relation between matter and antimatter? Are there really hidden dimensions in space?

To find the answers we must go where we've never been before—to the crucible of creation right after the Big Bang, a place in which the forces of nature reveal themselves in all their glory, a place in which the most fundamental particles pop in and out of existence, a place where we can see the primeval forces, fields, and particles that played crucial roles in the evolution of the universe and then disappeared from view. To do this we need to re-create the circumstances in which the forces and particles of nature were born.

Modern physics is the basis for all the progress we have made in understanding reality, perceiving the structure of the universe in which we live, and trying to come to grips with its laws, its bizarre phenomena, its particles, and its forces. And there is a curious and deep connection in nature: Whatever happens to tiny particles that are far smaller than what we can see with our eyes is *crucial* for what happens to the immensely large universe as a whole. "The interactions of particles *determine* the evolution of the universe," Nobel Laureate Jerome Friedman explained to me when I interviewed him about the great project of the LHC and its promise.[8]

These interactions among particles take many forms in their influence on the universe as a whole. As it turns out, a tiny anomaly in the way matter particles behave, as compared with the behavior of *antiparticles,* can explain why the universe exists at all, instead of being canceled out at birth. And the interactions of particles and *fields* (such as the electromagnetic field familiar to everyone who has played with magnets and iron shavings as a child) explain why the universe expanded after the Big Bang, and how it developed. The secrets of nature and the Earth, the Sun, the planets, outer space, and the distant stars and galaxies can all be traced to the mysteries of the behavior of nature's tiniest particles and the forces governing this behavior.

The four forces that control the behavior of the smallest particles have been established—gravity, electromagnetism, the weak nuclear force, and the strong nuclear force—but scientists want to understand the function of these four forces, and to know why they are of vastly varying strengths. For example, they want to understand why gravity is so weak compared with the other forces. You would never think of gravity as weak when you try to lift a heavy object, or to jump up in the air; events in our daily lives indicate to us that gravity is strong, at times too strong for comfort. But gravity is very weak in comparison with another force we are familiar with from everyday life— electromagnetism, the force responsible for the static electric shocks we may get when touching a metal doorknob after walking on a shaggy carpet when it's very dry; this is the same force that causes the motion of all electric motors, that makes the compass needle point north, and that enables radio and cell phone communications, as well as radar.

To see that gravity is much weaker than electromagnetism, you can perform a simple experiment. Place a paper clip on a desk and lower a small bar magnet toward it. When you get close enough, the paper clip will jump up in the air and stick to the magnet. This demonstrates that a very small source of the electromagnetic force, the bar magnet in your hand, can overcome the gravitational pull exerted on the paper clip by the *entire planet*. This shows you just how weak gravity is compared with the electromagnetic force, and physicists want to know why. Some have even hypothesized that the strength of the force of gravity is diluted because this force extends into some hidden dimensions of space that we cannot normally perceive. The LHC may shed light on this problem as well.

The example of the bar magnet winning against the gravitational pull of the entire planet, enabled by the fact that gravitation is much weaker than electromagnetism, can also be used to explain how the

LHC works. The paper clip has no choice: it is pulled up to the magnet instantaneously. The same happens with the protons accelerated by the LHC. The protons, with their positive electric charge, act like tiny paper clips: they cannot resist the power exerted on them by the giant magnets and radio frequency devices of the LHC. The magnets placed along the entire 16.5-mile circumference of the LHC constantly bend and focus the beams of protons, and the electromagnetic field of the radio frequency devices lends them greater and greater speed along the circular path underground. Gravity pulls down on the protons, but because it is so weak compared with electromagnetism, its effect on the protons is negligible.

When the protons finally crash against their counterparts traveling in the opposite direction, they release all their energy, thus re-creating the conditions that existed very shortly after the Big Bang. Right after the Big Bang, the temperature in the universe was immense. Particles of matter and antimatter were constantly being created and destroyed. We know that when antimatter meets normal matter, it immediately annihilates it, destroying itself in the process. So if matter and antimatter were created in equal amounts in the Big Bang, as scientists believe happened, how can we exist? Shortly after the Big Bang, for reasons we don't understand, matter is believed to have won over antimatter, leading to the existence of the matter in the universe we know, including stars, planets, and life. Some special-purpose experiments in the LHC, carried out in the LHCb detector, are aimed at solving this mystery. Other experiments, conducted at the ALICE detector, aim at understanding the nature of the "primordial soup" of particles that was our universe a small fraction of a second after the Big Bang.[9]

These questions about particles and forces and the nature and origin of all matter are the most important questions in physical science today—they are the fundamental questions about our existence, where we came from, what we are made of, and where we are going. To achieve the aim of providing answers, CERN was founded in 1954, and over the last two decades many of its efforts have been directed at creating the ultimate particle accelerator—the Large Hadron Collider. This is why the LHC is so important to science and worthy of the investment of so many billions of dollars and so much time and effort by the many thousands of dedicated scientists who work here and in collaborating institutions around the world.

But why do we need a particle accelerator to look for answers to our questions? Why not some other elaborate piece of equipment? The reason is that in order to discover the hidden particles of nature we need concentrated energy that can be turned into new matter. In the words of Nobel Laureate Frank Wilczek of MIT: "In the deep quantum world, to see something you must *create* it."[10] And as another Nobel Laureate, Yoichiro Nambu, puts it: "To discover new particles or investigate an unknown interaction, higher and higher energies are absolutely necessary. The 'energy equals mass' relationship limits the mass of the particle that can be created at a given energy; so if we have looked at all possible reactions with a given accelerator, we may say that the accelerator has fulfilled its fundamental purpose. Then we need an accelerator with the next higher energy—and so on."[11]

Since the accelerators that have preceded the LHC have all fulfilled their purposes long ago and have created, or revealed, all particles within their possible energy levels, the LHC is science's next big step. The highly anticipated experiments at the LHC are, in many ways, the focal point of all of modern physics—events in which all the theories developed over the centuries, all the previous experiments, and our collective imagination about nature all converge. It is also a place in which theory and experimentation merge as never before.

The LHC generates tremendous amounts of data on the results of its crashing protons. These data are analyzed by a state-of-the-art computing system called "the grid," an interconnected network of thousands of computers located in thirty-five countries. That CERN is the best place for computing innovations such as this one is supported by the fact that one of the most important developments to affect our daily lives took place here. In 1989, the World Wide Web was invented by a CERN scientist, Tim Berners-Lee, who was looking for a way to connect scientists and allow them to share their research results in the most efficient way. He was so successful in this task that the system of computer communication he devised has become one of the greatest inventions of all time.

Gaining a deep understanding of physics and cosmology is an important goal by itself, but potential discoveries may also have unexpected applications to everyday life. As the example of the World Wide Web has shown, an idea and the technology it spawns within a scientific setting such as CERN can have immense consequences for the technology and economy of the entire planet. If we didn't understand the laws of physics that govern radio waves, for example, then radio, television, the ability to land airplanes, cell phone conversations, and the Internet would not have been possible. In addition, new technologies such as computed tomography (CT) and positron emission tomography (PET) scanners, which are essential in modern medicine, were instigated by the science of particle accelerators, with major developments having been made at CERN. Medical science also uses accelerators to destroy tumor cells, and the operation of the LHC may teach us more about using this technology: how to focus particle beams to such precision that they would destroy cancer cells and leave healthy tissue unharmed. In this way the search for knowledge pursued at CERN and in similar high-energy laboratories around the world has value that goes far beyond our immediate scientific expectations.

One can imagine great new technologies that could well be born here when techniques for handling highly energetic proton beams, the like of which have never been created before, are understood to have other applications in our lives. The Medipix All Resolution System (MARS) scanner, for example, recently developed by researchers at the University of Canterbury in New Zealand, is a next-generation CT scanner based on technology pioneered at CERN. This new scanner, a revolutionary new "color X-ray machine" whose use promises to save many lives through early detection of tumors, has been developed as one of many spin-offs from particle research conducted as pure science.

The Terror of Black Holes

CERN was designed as a consortium of scientists from many countries working together to pursue knowledge. The scientists had to fight for what they believed in, and it has been a struggle to get governments to support this intensive, resource-hungry project in the name of science. Progress has been hampered by public fears that once the collider runs for some time at its colossal maximum energy of 14 teraelectron volts, seven times as powerful as anything seen inside an accelerator before (the Tevatron, at Fermilab in Illinois, generates just under 2 TeV), a black hole may be created. Such a monster, some people fear, could then grow and swallow the Earth. According to the Geneva newspaper *Le Temps*, a video on YouTube with a graphic depiction of a tiny black hole becoming progressively larger and finally engulfing the entire Earth, has been viewed by more than a million people.[12]

Even people who don't take the threat of black holes seriously recognize its prominence in the public psyche. Across the Atlantic from CERN, students at the Massachusetts Institute of Technology (MIT) devoted one of their famous pranks to poking fun at fears of a black hole. One night in late 2009 they sneaked into one of the main halls in the institute and hung from its domed ceiling a plastic spaceship with the letters CERN painted on its side, crushed out of shape and appearing to spiral into a black object hung from the center of the dome.

Oxford University physicist Alan Barr began a very technical presentation on dark matter at a physics conference in Boston in 2009 by showing a slide of the front page of the English newspaper *The Sun* from September 10, 2008—the day the LHC went on low-power testing for the first time. He said: "For those of you who don't know Britain, this is how the British public was introduced to the LHC."[13] The headline screamed that the LHC might produce a black hole that could swallow the Earth, followed by: "But don't panic, there is still enough time to try every position in the Kama Sutra."[14]

Could a black hole be produced at CERN? And would it swallow the Earth? Humor aside, scientists have considered this possibility carefully. The famous British theoretical physicist Stephen Hawking presumably hopes that a small black hole will appear at CERN—and then quickly evaporate. Such an occurrence could well confirm Hawking's famous mechanism for the evaporation of black holes (which for stellar-mass black holes takes eons but for a tiny one should take a fraction of a second), called "Hawking radiation," and could thus lead to a Nobel Prize. And a number of top physicists do believe that a tiny black hole could be produced by the machine, but scientists at CERN and elsewhere are quick to assure us that the probability that it would last is extremely small.[15] Still, none would say that the chance is zero. Everyone does agree that with the LHC we are entering completely uncharted waters in terms of the energy created by machine, and it is impossible to predict exactly what will happen once the accelerator is run continuously over a period of years.

Most physicists, of course, are much more interested in the exciting new physics and the potentially great discoveries about the nature of the universe—of which a tiny, quickly evaporating black hole might well be a part—that are opened up by the work of the LHC. On September 10, 2008, ignoring fears of a black hole that could swallow the Earth, CERN began its low-level testing of the LHC in preparation for finally making the protons collide. The protons were accelerated for twenty minutes in a very successful test run at well over 99 percent of the speed of light, but below the collider's maximum capacity. Representatives of the world's press organizations were invited to attend, and the CERN Control Center was filled with more than three hundred people. At that time, Manuela Cirilli, an attractive young woman with sparkling brown eyes and dark hair who holds a PhD in particle physics from the prestigious Scuola Normale Superiore in Pisa and who had worked with the ATLAS group since 2001, was helping with CERN's media relations. She told me what happened that day.

"It was actually very calm for a scientific experiment of this magnitude," she said, "but the chaos and excitement was in the headset over my ears." The media people were working on a webcast, people were moving into positions and speaking into microphones, and journalists were running around with their own equipment. But the scientists seemed unfazed—to a casual observer. Manuela saw deeper than that. She was looking at the scientists somewhat ruefully, perhaps envying their great moment, which had just arrived after so many years of preparation and expectation. Having gone over to media relations, she missed working in physics. She noticed that while the media were in a frenzy of jostling for positions to take in all the action, her fellow physicists at the controls of the giant machine followed a less frantic rhythm. "It was quiet, like a normal operation in which every person knows exactly what to do," she said.[16] But as a scientist, she understood that while the physicists seemed so serene doing their job, they were all fully aware of the huge responsibility placed on their shoulders.

Manning the controls of the LHC was Stefano Redaelli. "The night before," he confessed to me, "I couldn't sleep very well. I was too excited."[17] But once he came in to start work that day, the young physicist was relaxed and ready. He was eager to power up the collider to new record levels of energy. Manuela Cirilli recalled further: "I looked at Lyn Evans, and he was quiet, certainly, but you could see and feel the tension. And it was really great to see that tension relaxing after the beam went around the ring successfully." The day was a perfect success, "even better than we had thought!" Manuela exclaimed.[18]

Nine days later, the CERN scientists began more testing of the machine, which was to run at 5 TeV, a very high level of energy. Then, without any warning, the collider went dead. A problem had developed in a copper welding leading to one of the giant superconducting magnets, and the magnet quenched (meaning lost its superconductivity) and leaked its liquid helium; in a chain reaction that followed, 53 other magnets failed as well.

When even a tiny component of the system breaks down, the superconductivity of a magnet can be destroyed, and once this happens, the thick metal cables in the magnet's coil feel resistance. Resistance in turn creates heat. This unleashes a chain of increasing temperature and decreased conductivity until the magnet is destroyed. It can even cause a fire.

This is exactly what happened. When the resistance increased, all of a sudden sparks flew in the tunnel, causing an electrical fire that burned a hole in the magnet's outer steel layer. Then the supercold liquid helium leaked into the LHC tunnel, creating a domino effect that caused the other giant magnets to fail. It all took less than a minute, and the destruction was complete.

The culprit was a simple welding point that turned out to be too weak for the electric current it was to convey. No one could have predicted it. The LHC had been built from scratch—there has never been a prototype for such a powerful collider, so the scientists had to learn about the properties and capabilities of their creation as they went along. As a result of the accident, all the welding points of the LHC had to be redone and retested, and the task of repairing and rebuilding components for the machine took many months to complete. From this experience, the scientists learned that CERN needed to devise a new protective system for the magnets, called the quench protection system (QPS). This is a sophisticated electronic system that continuously monitors the magnets to make sure their temperatures do not rise and that superconductivity is maintained. It also monitors the 23,000 high-current joints in the entire machine to make sure they are functioning correctly.[19]

Unfortunately, a big celebration had been planned for the inauguration of the machine. European heads of state and many other dignitaries from around the world had been invited to "LHC Fest," but because the collider was now dead for the foreseeable future, the French president canceled his visit, sending a lower-level official. And the Italian government decided to send to CERN only its ambassador to the international organizations headed in nearby Geneva. For CERN, this was a public relations disaster. It was for this reason that when operations were restarted in late February 2010, no journalists were invited.

The accident was a traumatic event in the life of the collider. It made the scientists of CERN much more cautious about testing the limits of this giant of human ingenuity. They had to learn how to run the LHC much in the same way someone might learn how to drive a car without a teacher, since this machine had never been "driven" before and the limits of its performance were unknown. Also, implementing the QPS took a long time. Work progressed very slowly over the next year.

So slow was the progress, in fact, that the frequent delays and setbacks following the accident caused disappointment in the science establishment around the world: Scientists everywhere were losing patience, having expected miracles from the LHC. The *New York Times* even published an article by science editor Dennis Overbye that reported on a bizarre theory by two physicists that "the hypothesized Higgs boson, which physicists hope to produce with the collider, might be so abhorrent to nature that its creation would ripple backward through time and stop the collider before it could make one, like a time traveler who goes back in time to kill his grandfather."[20]

The long wait, and the problems CERN had had with getting the LHC program on track, was frustrating for all physicists. As Jack Gunion, a particle physicist from California, put it at a high-level physics conference held a few months before CERN was able to resume the activity of the hobbled LHC: "We're all going crazy waiting for the Higgs!"[21]

And then, seemingly overnight, the collider surprised everyone and came back to life with a vengeance. It was fully healed from the breakdown of September 2008, undeterred by fears that it could destroy the

Earth, proof positive that it was hearing no voices from the future. The machine was up again and roaring. Every few days, a new world record was achieved at CERN: A speed record, an energy record for particles in an accelerator, and then the LHC's own previous records were successively broken by the collider.

At 12:44 a.m. on November 30, 2009, with Stefano Redaelli at its controls and Lyn Evans somewhere behind, the LHC broke the world energy record set by Fermilab some years earlier, by accelerating protons to 2.36 TeV total. Two weeks later, protons were made to crash at this energy level. Bottles of expensive champagne were popped open at CERN Control Center, and everyone celebrated. "Isn't this the greatest machine ever?" people asked one another joyfully.[22] Then, during the day, the various scientific collaborations at CERN—CMS, ATLAS, ALICE, and LHCb—each in turn sent people to the CERN Control Center with their own champagne bottles to celebrate records for collision energies broken by each of the four teams, and the celebration was complete.

"This was just amazing," recalled Redaelli, by then quite tired, having stayed up a night and a day. "It was a complete team effort," he said, "Everyone contributed to it."[23] As 2009 wound down to a close, on December 16, CERN sent an exuberant "tweet" to its thousands of Twitter followers worldwide: "The LHC's first run came to a close at 18:03 CET [Central European Time] today after a successful series of 'firsts.' Back in business next year."[24]

Chapter 2

The LHC and Our Age-Old Quest to Understand the Structure of the Universe

The ancient Greeks marveled at the universe around them. Driven by desire to unlock the secrets of creation, they set out to discover the basic building blocks of matter. The fifth-century B.C.E. Greek philosopher Democritus proposed the theory that matter was composed of tiny unseen particles resembling the atoms we know from modern science. And in the same age of classical Greece, the mathematician Eratosthenes, who lived in Egypt, devised an ingenious method, using measurements of the angle to the sun from two locations a known distance apart, to estimate the circumference of the Earth to a surprising level of accuracy—very close to today's known value of around 25,000 miles. This dual search for knowledge, both in the realm of the small and that of the large, continued throughout history.

Galileo, "the father of physics," studied small objects and their trajectories—perhaps even by dropping them from the Leaning Tower of Pisa, as the legend goes. At the same time, he was the first person to turn a telescope toward the heavens, where in 1609–1610 he discovered the four largest moons of Jupiter—the "Galilean satellites," named after him.

Guido Tonelli, who is from Pisa, where Galileo performed his famous experiments, told me that he felt a strong kinship with the great scientist while doing his own calculations related to the particle interactions measured by the CMS collaboration. "I realized that I was using the same technique that Galileo had used four centuries ago when he correlated the time measurements against the heights from which objects were rolled along an inclined plane," he said. "I was using the same idea when correlating particle energy measurements and time."[1] He explained how Galileo "weighed" time by placing on a scale a container into which water was dripping at a constant rate in an experiment in which time was a key variable. While our technology has changed so much in four hundred years, human ingenuity has not.

Galileo's contemporary Johannes Kepler derived the laws of planetary motion based on the observations of Tycho Brahe, thus bringing us an understanding of the workings of the solar system. And Isaac Newton, born in 1643—a year after Galileo's death—studied how gravity works on Earth and at the same time applied his theory of gravitation to the larger universe, explaining Kepler's laws. Using his own laws of gravity, derived from falling objects on earth, Newton was able to deduce that the force of gravity was making the Moon "fall continuously" to Earth—that is, revolve around us, forever attracted to Earth by the force of universal gravitation. Building on the foundational work of Galileo, Newton thus established an important connection between the large and the small in the universe through the mysterious force of gravity.

In the words of George Smoot, who won a Nobel Prize in Physics in 2006 for leading work in cosmology, "In a sense, physics and astronomy became one subject under Galileo's genius. It was the grandest marriage of two physical sciences until the late twentieth century, when cosmology and quantum particle physics began to merge."[2]

Today we recognize a unity of science as we study both the small and the large in our modern quest for knowledge. Particle physics and cosmology have become allied fields, and the links between them keep getting stronger. Study of the behavior of elementary particles leads to an understanding of the early universe, and equally, what cosmologists and astrophysicists learn about the large universe around us also teaches us things about the nature of the small particles that make up all matter in the universe. ·

In order to understand the basic facts of modern physics that underlie much of the science being advanced by the LHC, we need to take a closer look at the theories that have been proposed to explain nature.

Relativity and Quantum Mechanics

Newton's classical theory of mechanics works very well in describing the behavior of reasonably large objects. But as soon as we enter the realm of the atom and its constituents, things change drastically: The rules of behavior we know from our everyday world no longer apply. Equally, when speeds of objects approach that of light, or masses become extremely large, the old physics doesn't work anymore either. In these realms, *time itself* is not what we think: It can slow down. And space can become warped and objects malleable.

Between 1900 and 1930, physics changed unrecognizably because of a dual revolution: first, the emergence of quantum mechanics, launched by the work of Max Planck in 1900 and extended over the next three decades by a group of brilliant young physicists; and second, the birth of the special theory of relativity in 1905 and the general theory of relativity in 1915, both achieved through the work of one man, Albert Einstein. In 1905, the unknown twenty-six-year-old stunned the world of science by presenting not one but *four* different papers, each of which transformed how we view the physical world. One of them, on the photoelectric effect, showed that light can be viewed as composed of particles (this paper was the only achievement to win Einstein the Nobel Prize). A second paper, on molecular motion, outlined a major breakthrough that demonstrated the existence of atoms. In the third paper, Einstein presented his special theory of relativity, in which he showed that time is not constant, as had been thought for millennia, but slows down at very high speeds (those approaching the speed of light); that length contracts at such high speeds; and that the speed of light is the maximum speed limit in the universe.[3]

The fourth paper, an outcome of Einstein's work on special relativity, presented his discovery of the equivalence of energy and mass, stated by his famous equation $E = mc^2$. This result opened a whole new world to be explored. As Einstein himself put it in an interview in the documentary *Atomic Physics*, "It follows from the special theory of relativity that mass and energy are both but different manifestations of the same thing."[4] Everything that scientists at CERN and other particle accelerator laboratories do is based on Einstein's incredibly powerful formula.

The particles that are smashed together inside the LHC or other particle accelerators have two kinds of energy stored inside them. One is the rest energy, the energy of a particle that is not moving at all. For example, the rest energy of an electron, based only on its mass, as if it were not moving at all, is 0.511 megaelectron volts (MeV), or about half a million electron volts. This is the "natural" energy stored inside an electron, without its energy of motion. It is the energy that we would obtain if we destroyed one motionless electron, turning all its mass into energy.

When we accelerate a particle, or it is naturally moving, as electrons usually do, it is seen to possess a *second* piece of energy. This is its energy of motion, called *kinetic* energy. The total amount of energy provided by a particle that is completely destroyed inside an accelerator is the sum of the two energies: the rest energy plus the kinetic energy (although it is not computed as a sum but is based on a more involved formula that accounts for both kinds of energy).[5] Now comes the most amazing part of this energy-mass game. Inside the LHC, right where the protons crash and are disintegrated, a certain amount of energy is released. Now, according to Einstein's famous formula, energy is equivalent to mass. So the energy produced when the protons crash is now available for the formation of other particles!

But which particles can be created here? In theory, what can be created is any set of particles whose total energy is equal to the energy produced here—this is in keeping with a concept called the *conservation of energy*: energy-mass cannot be created or destroyed, so the total you started with must equal the total you end up with. Energy-mass can only change its form: Some of the energy can turn into mass, and some mass can change into energy, but the total mass-energy stays the same. (Other conservation laws in physics, such as the conservation of electric charge, must also be obeyed by these reactions inside the collider.) To put the energy conservation idea in perspective, we know that the energy of an electron at rest is 0.511 MeV, but the total energy produced by the crash of two protons flying at full LHC speeds (each proton at 99.9999991 percent of the speed of light) is 14 TeV, a much larger amount of energy—millions of times more. So certainly lots of electrons can be produced. But since the energy released by LHC collisions is so high, much heavier particles than electrons can also be created. And this is why the LHC was built—to create, or make appear, particles we've never seen before. This can happen as long as their total mass-energy is within the TeV range available to the LHC. This should hold true for the Higgs boson, some supersymmetric particles, or other particles whose existence we have not even imagined.

In 1900, the German physicist Max Planck deduced that energy can only come in discrete "packages," called *quanta,* after studying radiation produced by a heated object (called blackbody radiation). This finding launched the field of quantum mechanics—the area of physics that studies the behavior of small particles. Einstein's paper on the photoelectric effect, published five years later, explained some of the mystery of how the particles of light, the photons, interact with matter. His idea constituted the first part of the later concept of particle-wave *duality,* because it established that light—which until then was viewed as a wave—also possesses some properties of a particle. Later, the French prince Louis de Broglie further established particle-wave duality, which asserts that all small particles (and some that are not so small, as we will see) possess both the properties of a wave—such as diffraction and interference—*and* the properties of a particle.

Ultimately, the model of the atom we have today is based on quantum theory. According to its rules, including Heisenberg's famous uncertainty principle, there are no precisely known locations, velocities, and other parameters of particles, but only sets of probabilities. Electrons, for example, do not have well-defined positions in their orbits around atomic nuclei at any given moment in time, unlike the planets in their revolution around the Sun.

Quantum theory was developed in part because the solar-system-like model of the atom raised a conundrum: Why did the electrons not fall into the nucleus and combine with it? Clearly, opposite charges attract each other, so there had to be something in the atom to keep the negatively charged electrons from tumbling into the positively charged nucleus. The answer to this question was provided by the Danish physicist and quantum pioneer Niels Bohr, who constructed a more sophisticated model of the atom than had been known before, founded on the emerging principles of quantum mechanics.

In 1913, Bohr presented a model based on Planck's earlier discovery of quanta. Bohr's hypothesis was that the electrons in an atom orbited the nucleus only in orbits with precisely determined levels of energy. According to Bohr, the electron in an atom can only be in an orbit with a given radius (and hence a given energy level) and can drop down only to another orbit with a specific energy level—not to an arbitrary orbit we might choose. Only *quantized* levels of energy were allowed.

Once his model of the atom received acclaim in the scientific community, Bohr became a prominent physicist, and the Carlson brewing company in Copenhagen financed an institute of theoretical physics to be headed by him. The Danish royal family also invested heavily in physics. In addition to salaries at Bohr's institute, it also supported the many visiting physicists who came here from around the world. These scientists included other quantum pioneers, such as the Austrian physicists Erwin Schrödinger and Wolfgang Pauli, the German physicist

Werner Heisenberg, the English physicist Paul A. M. Dirac, and many others. Through the work of these great minds, quantum physics developed far beyond its rudimentary beginnings under Planck and Bohr.

The break from the old physics also took another direction. Albert Einstein finished his greatest creation, the general theory of relativity, in November 1915 and published it in 1916. It was the crowning glory of Einstein's life work. Because it joins his special theory of relativity with Newton's mechanics, the result is a comprehensive theory of gravity, which has wide-ranging implications. It allowed Einstein to explain the behavior of the force of gravity in a much wider range of contexts: speeds approaching that of light and very massive objects. In fact, even to explain the orbit of Mercury, the closest planet to the Sun, requires general relativity. Newton's laws fail to account for an anomaly in this orbit due to Mercury's nearness to the massive Sun. General relativity has also led to the prediction of the existence of black holes and many other phenomena.

A revolutionary idea in general relativity is that mass creates *curvature* of space-time. For example, space around the Sun is curved, and this spatial curvature is what makes the planets revolve around it. These planets, in turn, also curve space around them, which is why the Moon revolves around the Earth. This was a more precise explanation of the Moon's movements than Newton had been able to provide two centuries earlier using his theory of gravitation. But because gravity is the weakest of the four forces of nature, forty orders of magnitude weaker than electromagnetism, it can be completely ignored in particle physics. Small particles can exert on each other—in any measurable sense—only the other three forces: electromagnetism, the weak force, and the strong force. Which of the forces a particle "feels" depends on the kind of particle.

A proton, for example, can feel the electromagnetic pull of electrons in the atom, the weak force acting within the confines of a nucleus and its components, and the strong force that holds it inside the nucleus and keeps the quarks inside it from breaking loose. But it cannot feel in any measurable way the negligibly small gravitational pull of an electron or of another proton, or even of a giant molecule nearby. And the tiniest fermion we will meet, the neutrino, which has no electric charge and extremely little mass, feels only the effects of the weak force (and gravity to an immeasurably small degree). Particle physics, as studied experimentally at CERN and elsewhere, is based on the Standard Model of particle physics, which does not include the effects of gravity. Hence it is not tied in with Einstein's *general* theory of relativity, which is a theory of gravity.

The Holy Grail of physics is a "theory of everything," a single model using one or several equations that will "capture" the effects of *all* the forces of nature, including the very weak force of gravity. As such, the "theory of everything" must be an amalgam of both quantum theory and Einstein's general theory of relativity. Any experimental progress in this direction during our time (if it is at all possible) will likely come through discoveries at the LHC, where the highest energy level we can now reach is available.

Einstein's own pioneering work left him puzzled about the quantum world he helped discover. Everything Einstein believed about nature ran contrary to the principles of the quantum theory, mainly because he did not believe in the role of chance in the behavior of the particles of nature and famously exclaimed that he did not believe that God plays dice.[6] Equally, he did not believe in nonlocality, the principle that events in very separate locations of space can affect each other without sending a message at the speed of light. Nonlocality is a direct outcome of quantum theory, exhibited by a phenomenon called entanglement.[7]

These contradictions left Einstein frustrated and combative. He constantly attacked quantum ideas in letters and papers and in conversations with the physicists who were continuing to develop quantum theory. He always looked for—and being Einstein, he often found—holes in the arguments, which he used in his assaults on the emerging quantum view of nature. Einstein's disagreement with quantum principles would turn into a deep dissatisfaction just as the young quantum pioneers Heisenberg, Pauli, and Dirac, as well as the older Schrödinger and Bohr, would push the theory to greater heights.

The first physicist to build on the ideas of Bohr and Planck was twenty-four-year-old Werner Heisenberg. He discovered a crucially important concept of quantum physics—the idea that a particle's momentum and position cannot *both* be measured with precision, regardless of the accuracy of the apparatus. If one is known with precision, the other necessarily entails some uncertainty. This came to be known as Heisenberg's uncertainty principle. Heisenberg showed that this quality is endemic to behavior in the microworld of atoms, molecules, electrons, protons, neutrons, and the like.

After Heisenberg's work appeared, Wolfgang Pauli used the new quantum ideas to solve the energy states of the hydrogen atom. That is, he was able to use quantum mechanics to derive something that could be tested and measured: the behavior of the simplest atom, hydrogen. Pauli was born in Vienna in 1900 but spent most of his adult life in Switzerland, where he was a professor of physics at the Eidgenössiche Technische Hochschule (ETH), the Swiss Federal Institute of Technology in Zurich. Pauli was a fierce anti-Nazi, and living in Switzerland thus ensured his survival over the period Hitler was in power and controlling most of the rest of Europe. A large man with a resonating laugh and a pleasant disposition, Pauli is mostly remembered for the Pauli exclusion principle—if there are two electrons in the same orbit, they must spin in opposite directions—and for discovering the existence of the neutrino through theoretical analysis.

Pauli seems to have had a difficult time with his personal relationships and often suffered from anxiety. As fate would have it, Pauli met the famous psychology pioneer Carl Jung. Jung analyzed many of Pauli's dreams, leading to interesting hypotheses about the relation between quantum physics and the mind.

In 1926, the Austrian physicist Erwin Schrödinger developed a complementary approach to Heisenberg's quantum theory using a wave equation later named after him, thus extending the idea of the particle-wave duality proposed a year earlier by Louis de Broglie. Schrödinger's equation uses the wave properties of a particle and can only be solved in closed form in simple cases; for more complicated quantum systems a computer is required. The solutions of Schrödinger's equation have a peculiar property. They allow a *superposition* of states. In our everyday world, a light switch can be either "on" or "off"—not both. But in the weird quantum world, an entity can be in both states, "on" and "off," at the same time, which means that a quantum particle can be both here *and* there at the same time, rather than here *or* there. To exemplify this bizarre quantum phenomenon, Erwin Schrödinger devised a thought experiment now widely known as "Schrödinger's Cat."

Schrödinger's Cat works in the following way: A cat is placed in a closed box. Inside the box is a vial of cyanide and a hammer attached to an electric mechanism triggered by a speck of radioactive material. When an atom in the radioactive element disintegrates, it triggers the mechanism, the hammer breaks the vial, the cyanide is released, and the cat dies. The radioactive disintegration is a quantum event, governed by the laws of quantum mechanics. Schrödinger asked the following question: If we don't know whether the radioactive disintegration has taken place, is the cat alive or dead? His answer was that, in keeping with the laws of quantum theory, the cat is both alive *and* dead at the same time.

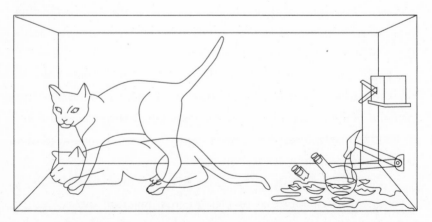

It should be noted that the Schrödinger's Cat paradigm is an attempt to impose quantum conditions on a macrosystem, so it has many flaws. But the idea nicely illustrates the weirdness of quantum mechanics. We will use this principle throughout the book. For example, the direction of spin of an electron is a *superposition* of two possibilities: up and down. It is this idea of superposition that will lead us to continuous symmetries, which are so important in physics today.

In 2004, Schrödinger's daughter, Ruth Braunizer, invited me to visit her in her father's house in Alpbach, Tyrol, in the Austrian Alps. We drove high up the mountain and finally reached a lovely alpine house with a sign in front: DAS SCHRÖDINGER HAUS. We were welcomed by the cat—a descendant of a famous ancestor but still a Schrödinger cat—a small, gray feline that snuggled up to me and purred: It was very much alive.

In the late afternoon, I walked down the wooded path in front of the house to Erwin Schrödinger's grave. On the commemorative cross Ruth had placed a metal plate with a poem her father had written in 1942. I like the way the poem captures some of the flavor of quantum mechanics—and Schrödinger's life philosophy, as inspired perhaps by quantum ideas. As the last rays of the setting sun between the mountains to the west lit up the grave, I read the German words (here translated into English by Ruth's husband, Arnulf Braunizer):

What is, is not because we feel it
And is not naught because we cease to feel it.
Because it is, we are, and are forever.
So all being is one only being.
And that it lasts when one is dead
Means that one has not ceased to be.

E. SCHRÖDINGER, 1942

Cosmology, Symmetry, and the Laws of Conservation

Our understanding of the evolution of the universe as a whole was enhanced in the early 1980s by the groundbreaking work of Alan Guth of MIT, a particle physicist turned leading cosmologist. Working at the Stanford Linear Accelerator Center (SLAC) in California, Guth made a major discovery about the evolution of the universe: He hypothesized the phenomenon of *inflation.*

According to Guth's theory of the inflationary universe, right after the Big Bang our universe experienced a period of intense, exponential expansion in size. Within a fraction of a second, the universe grew from the size of a subatomic particle to that of a marble.[8] When inflation stopped, the universe continued to grow at a more moderate pace, and it expands today. In our era, however, this expansion is accelerating, as has been determined by astronomical studies of the rates of recession of distant galaxies concluded in 1998. One big mystery in cosmology is the nature of the mysterious force that accelerates the expansion of the universe. It is called "dark energy," as distinct from "dark matter," which is another unsolved problem in cosmology.

Mathematically, space overall seems to have a Euclidean geometry, characterized by straight lines rather than curves: The universe at large does not appear to have significant curvature except around massive objects, where curvature is dictated by Einstein's general theory of relativity. The initial, exponential era of growth after the Big Bang, the inflationary phase, caused the universe to become "flat." These findings have been confirmed through successive studies of the microwave background radiation in space, obtained from satellite data over more than a decade.

Cosmology and astronomy have thus thrown particle physicists a huge problem to solve—just when physicists were getting complacent about their trusted models. And dark matter is an equally big, older problem. When in the 1930s Fritz Zwicky realized through observations of galaxies that they were held together far more strongly than would have been predicted based on all the matter that can be observed, he concluded that the universe must be filled with large amounts of dark matter. This missing matter is presumably composed of tiny particles that we have never seen before.

The LHC may help solve this mystery by discovering some of these particles. In the words of Alan Guth: "Within the next five to ten years we expect to know how much dark matter there is. . . . We will probably also know whether the complex fabric of cosmic structure can be explained, in detail, as the result of random quantum processes that took place during the first trillionth of a second in the history of the universe."[9]

Guth has explained that the LHC is a time machine that allows us to study the universe as it was a fantastically short time (5×10^{-15} seconds, or five thousand-trillionths of a second) after the Big Bang. By comparison, the first cyclotron, built at Berkeley in 1930, could give us a glimpse of the universe as it was 200 seconds after the Big Bang; the Cosmotron, built in 1952 at the Brookhaven National Laboratory on Long Island, brought us to 3×10^{-8} seconds (three hundred-millionths of a second) after the Big Bang; and the Tevatron, built at Fermilab in 1987, brought us to 2×10^{-13} seconds (two ten-trillionths of a second) after the Big Bang.[10]

The seeds of the formation of galaxies in the universe were discovered by a team headed by George Smoot. These scientists were the first to demonstrate how galaxies formed from the clumping of matter that took place shortly after the Big Bang. This early matter of the universe came from an extremely hot and dense "primordial soup" of elementary particles. From it, nuclei of simple atoms, mostly hydrogen, helium, and some lithium, were formed. Later, after early generations of stars had burned their hydrogen and helium fuels in fusion reactions and died, larger atoms from these processes were spewed into space and now form the atoms and molecules of the matter we know.

Like Alan Guth, Smoot also began his career as a particle physicist and later changed to cosmology. There Smoot made his mark on science by deciphering the first microwave radiation data set from space. Recalling his big moment of discovery in 1991, when he and his team found the primordial seeds that led to the formation of the galaxies, Smoot said: "It's easy to be fooled, so you have to be careful not to fool yourself and see in your data what you expect to see, rather than letting the data tell their own story." But after carefully repeated analyses, the data indeed showed clearly the seeds of the matter that later grew to become galaxies.[11] This news caused a great sensation in cosmology.

The seeds of the galaxies are the early clumps of matter formed when the soup of elementary particles coalesced into protons and neutrons, forming nuclei. The nuclei later combined with electrons to form simple atoms. To these, much later, were added more complex, larger atoms after the early stars had burned their fuels.

Atoms are the smallest constituents of pure elements: hydrogen, helium, lithium, carbon, oxygen, nitrogen, iron, copper, and so on. At its core, each atom has a nucleus, which is very small when compared with the whole atom and very dense; it contains electrically positive components, the protons, and also electrically neutral components, the neutrons. The rest of the atom is made up of the electrically negative electrons, which orbit the nucleus.

These orbits and the empty space they encapsulate form most of

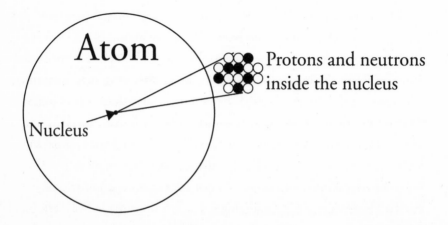

the volume of the atom. The volume of the nucleus itself is much, much smaller than that taken up by the atom as a whole. If an atom were the size of a bus, then the nucleus would be the size of the dot on the letter "i" in a newspaper read by a person on the bus.[12]

The proton is much more massive than the electron: One proton weighs as much as 1,836 electrons.[13] (Yet in our example it still fits inside the dot on the "i," because it sits inside the nucleus, and the electrons that surround the nucleus enclose far more space than the size of the very dense nucleus.) The neutron weighs only slightly more than the proton. Protons and neutrons are packed together inside the nucleus. The more of them there are in the nucleus, the heavier the element. Hydrogen has only one proton as its nucleus; the helium nucleus has two protons and two neutrons; and so on, as the elements get heavier.

But science went beyond the study of the atom, its nucleus, and its electrons to an analysis of the tiny components of the nucleus: the quarks that grouped in threes as the primordial soup cooled down after the Big Bang and created the protons and neutrons, and other elementary particles. Today we are on a quest to understand the behavior of these tiny particles and the fields and forces that govern their behavior. The way to plumb the depth of the atom experimentally was opened by particle accelerators, but the physical theory needed for understanding the experimental results required deep mathematics.

Galileo said: "The book of nature is written in the language of mathematics." And the new physics requires far more complicated mathematics than Galileo could have imagined. These are special kinds of mathematics, and the need to use them has brought great mathematicians into the field of physics, hoping to help shed light on the mathematical mysteries of the physical universe. In his keynote presentation at the conference "Perspectives in Mathematics and Physics," held at MIT in the spring of 2009, Fields medalist Sir Michael Atiyah of Oxford University, one of the greatest living mathematicians, described the five problems facing physics today:

Can we verify the Standard Model and find the Higgs boson?

What is the nature of the mysterious "dark matter"?

What is the "dark energy" that seems to permeate space?

Can we solve the mystery of matter versus antimatter?

Can we attain a deeper understanding of quantum mechanics?

To solve the problems Atiyah has described and hopefully someday to produce a final theory of physics, mathematics and physics must come together more closely than ever before.

Atiyah recalled that when he first came to MIT in the 1950s, mathematicians and physicists shared a building. "There was a door between the two departments, mathematics and physics, and that door was always locked," he recalled. "So I asked the physicists why the door was locked, and they told me that they had a new carpet and didn't want the mathematicians to come in and walk on it with their dirty boots."[14] Since then, Atiyah noted, much better relations have been forged between mathematics and physics, an important trend that must continue if we hope to someday arrive at a final theory.

One particular concept in mathematics, the idea of *symmetry*, which is studied through a branch of mathematics called *group theory*, plays an important role in modern physics. Symmetry is so intuitive and natural that our ability to perceive it seems almost hardwired into our brains. A face has a two-sided symmetry, for example, and we know that infants can recognize the symmetric features of a human face even a few weeks after birth. A starfish on the beach has a fivefold symmetry, which also appeals to our shape perception and sense of beauty in nature. And the human-made stop sign at the corner has an eightfold symmetry.

But the concept of symmetry goes much deeper. There are symmetries that do not appear in the usual space we see around us, but rather in more abstract settings, and such symmetries are essential to the understanding of theoretical particle physics. These ideas were pioneered by a very young nineteenth-century French mathematician, Évariste Galois, who died in a senseless duel over a woman at age twenty. Before he died in 1832, Galois introduced the mathematical concept of a *group,* which is our basic algebraic tool for understanding symmetry. Galois began his study while still a high school student. His ideas were so advanced for his time that no one understood them, and as a consequence he could not even gain acceptance to a good university, which led to frustration, naive involvement in revolutionary politics, and ultimately the tragic duel.[15]

Let's look at some symmetries. A face looks the same only if it is flipped over so that left becomes right and right becomes left. An equilateral triangle looks the same if it is flipped over along an axis through the center of any side or if rotated one-third or two-thirds of the way around in either a clockwise or a counterclockwise direction.

But a circle looks the same when rotated by *any* angle whatsoever: Whether rotated a tiny angle of a degree or less, or an angle of 90 degrees, or 223 degrees, it will always look the same after rotation. We say that a circle has a *continuous* symmetry and that the group of transformations (changes we can make to it that will leave it the same) is a *continuous* group. Such a group is called a Lie (pronounced "lee") group, named after the Norwegian mathematician Sophus Lie, who in the nineteenth century extended Galois's ideas about groups to continuous symmetries. Continuous symmetries are the ones most often used in particle physics.

In physics, we use a number of continuous groups to model physical processes. Whenever a continuous symmetry exists and can be modeled by a Lie group—that is, by a group of transformations that are continuous, such as the group of all possible rotations of a circle—we know that we have learned something important.

Nobel Laureate Sheldon Glashow of Boston University had this to say about symmetry:

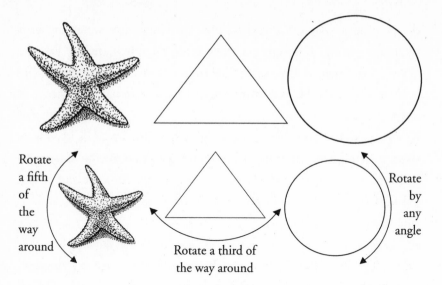

Rotate a fifth of the way around

Rotate a third of the way around

Rotate by any angle

Symmetry is always present in nature. Physicists for hundreds of years have been aware of the symmetries of crystals. In fact, it was the symmetries, the beauty of crystals and forms, which led some people to believe in the existence of atoms. But sometimes symmetries are there which are not so evident. This is the case, for example, with the neutron and the proton, two particles that are very similar in some ways but very different in others. This is the case of an apparent symmetry, an approximate symmetry, which is in fact broken in nature. Other types of symmetries are completely hidden or destroyed apparently by nature.[16]

To understand this idea, let's go back to the starfish. The "idea of a starfish" is something that has a perfect fivefold symmetry: You can rotate an idealized starfish in either direction, clockwise or counter-clockwise, a fifth of the way around, and it still looks exactly the same. But a living starfish you find on the beach is never perfect. Often one limb is smaller than the others, sometimes more than one. Obviously, the imperfections have developed while the organism was growing, and they could be genetically or environmentally caused. But you can think of a starfish as something that on the drawing board was perfectly symmetrical, but its symmetry was "broken" by nature.

Some symmetries are not broken, but hidden: Until the first time that humans saw the inside of the body (certainly still in prehistoric times), they didn't know that we have two kidneys, symmetrically placed on either side of the spine. This symmetry is hidden by nature. And you can think of the symmetry of the human lungs as one that was broken by nature: The "drawing board" idea is of two symmetrical lungs, but there was need for a good chunk of space for the heart and the arteries emanating from it, so the symmetry of the lungs had to be destroyed. The left lung doesn't look like the fuller right one.

The mathematical idea of continuous symmetry comes into physics in a very fundamental way through a result known as Noether's theorem. Emmy Noether was a German Jewish mathematician working at the University of Göttingen in the second decade of the twentieth century, where she proved that any continuous symmetry that can be modeled through standard tools in physics implies a *conservation law*.

The idea of a conserved quantity is paramount in physics. In a closed system, energy is conserved: It cannot be created or destroyed. As we know, energy can turn into mass and mass into energy, but the total amount of mass and energy in a closed system must stay the same. Equally, we know that electric charge is conserved, and thus its total amount in a closed system cannot change; electric charge cannot be created or destroyed. The same holds true for momentum. As a simple example of the conservation of momentum, a spaceship must apply its jets to slow down because momentum (in the vacuum of space) is conserved, and the spaceship would otherwise continue to move ahead. An example of the conservation of charge can be seen when you rub against a fluffy surface and then touch someone: that person will get a shock of static electricity because the charge in your body is conserved—it can't just disappear, so it goes to the person you touch.

Conservation laws are very important in physics, and they are linked to symmetries through Noether's powerful theorem.

The connection between symmetry and conservation laws is an important tool for the theoretical physicist. Once a symmetry is discovered, a conservation law must be lurking around the corner. This conservation law can in turn explain particle interactions, because it says that some quantity (whatever the conserved quantity may be—electric charge, energy, momentum) has to be completely accounted for. If the total amount after an interaction doesn't add up to the total amount before the interaction, something is missing. One beautiful historical example of the use of a conservation law was the Austrian physicist Wolfgang Pauli's prediction in 1930 of the existence of the neutrino. Pauli knew that energy must be conserved in nature, and he noticed that in a certain nuclear process, called beta decay, a small amount of energy was missing. He concluded that this missing energy was escaping from the decay process in the form of an additional particle, thus predicting the existence of the neutrino!

Knowing that some physical quantity is conserved also tells the physicist that an associated symmetry of nature must be hiding somewhere, and finding that symmetry may have great value in explaining the nature of the phenomenon under study, which can then lead to new theories about the world. Symmetries and laws of conservation, as well as special relativity and quantum theory, are all essential elements in the great experiment conducted inside the Large Hadron Collider.

Chapter 3

A Place Called CERN

I arrived at the Geneva airport for my first visit to CERN at 10:35 a.m. on April 2, 2009, after an hour-long flight from Paris. I almost didn't make it. There had been an electrical problem with the Paris RER (Réseau Express Régional) rail system, and my train had stopped in the north of the city, releasing hundreds of people to grab taxis or crowd their way onto buses to continue their journeys. It was an impossible situation, and I was about to give up on my trip when I was able to convince an indignant Parisian woman to allow me to share her cab to Charles De Gaulle Airport. Once there, I ran as fast as I could to the terminal, making it through the airplane door just before it closed.

Having finally arrived in Geneva, I met my contact, Dr. Paolo Petagna, a tall, young-looking Italian physicist from Livorno. After a few moments of conversation, we got into his car and drove west toward CERN. First we passed through neighborhoods with high-rise apartment buildings, which then gave way to a landscape of open fields and small villages. I noticed an unusual number of high-power lines overhead. "I guess we can't complain about being surrounded by so many high-power electric lines all over here—given the level of consumption we have," Paolo laughed, in response to my question.

We reached the gates of the facility and the security checkpoint. Paolo flashed his badge, and we were allowed to enter the place called

CERN. The acronym originally stood for Conseil Européen pour la Recherche Nucléaire (European Council for Nuclear Research), the name of the provisional council established for setting up the laboratory. This name was later changed to Organisation Européen pour la Recherche Nucléaire (European Organization for Nuclear Research), but the original acronym has been kept.

CERN's main site is a large research campus with long, white, multistoried buildings facing streets named after famous scientists. We drove past Route A. Einstein, Route N. Bohr, and Route J. Bell—the last one named after CERN's own quantum theorist whose work has led to our understanding of the bizarre concept of quantum entanglement.[1] After parking, we entered the first floor of the main building, with its large cafeteria and adjoining dining areas. It was coffee time when we arrived, and many of the scientists here, a true international mixture of people from all over the world, were coming in, talking to friends, getting coffee and pastries, and milling about. I heard a mélange of languages and saw people dressed in a variety of styles, including traditional Indian and African dress. On a bulletin board I noticed postings in both English and French.

Paolo anticipated my question, and explained: "When you come here, from wherever you arrive, you must choose one of the two languages as your language of communication at CERN," he said.

"And you chose French?" I asked, having heard him speak it fluently everywhere.

"No," he said with a smile. "I chose English. When I came here twelve years ago, my English was good, but my French was not. Now, after all these years, my French is much better and I feel more comfortable speaking it. But I remained listed as using English for all internal CERN communications. The English spoken here is a sort of standard 'International English.'"

"The language you hear at international scientific conferences all over the world," I said.

"That's it," he agreed.[2]

French is still more commonly used here, and it is "more official" than the nominally equal English—in part because CERN is located in the middle of a French-speaking area. The eastern part of the CERN complex, including the headquarters, in which we now were, lies in the French-speaking Swiss canton of Geneva, and the western part of CERN, where most of the LHC is located, lies in the Rhône-Alpes region of France.

As we sat down to drink our coffee, Paolo recalled: "When I first arrived here, I experienced a culture shock. The place is completely different from anything I had ever experienced in my life before." I looked at him expectantly, and he explained: "You see, normally, you would think that people with big egos compete with each other fiercely, and maybe they view other scientists as their enemies. But not here. What was so surprising to me was to see that the people working here have a unique ability to define common goals. This is something you don't see very often outside CERN."

I learned that what CERN has is something I would call "healthy competition"—everyone obviously wants to be the one to make a big discovery, but they go after their goals in a cooperative way. I saw none of the ruthlessness I had seen in so many other places. The scientists, engineers, and workers were uniformly friendly and helpful, and clearly the relations among them were exceptionally good. Still, the positive aspect of competitiveness is nurtured here too. The two main multipurpose detectors, ATLAS and CMS, are run by highly competitive teams. These two detectors are looking for the same kinds of phenomena: a Higgs boson, dark matter candidates, extra dimensions of space, supersymmetric partners, strings, and other discoveries. But the *way* the two groups operate is different and separate from each other.

Each of the two teams has designed its own detecting apparatus, its own construction design, and its own scientific methods. And each team hopes to beat the other to great discoveries. As a member of CMS, Paolo Petagna later showed me the CMS detector and explained how it works. In fairness to his competitors, he also arranged for me to meet the spokesperson (CERN-speak for "director") of ATLAS, Dr. Fabiola Gianotti, a fellow Italian. "It's useful to have competition," Petagna explained, "and we also have a strong unification of the two teams—in the sense that we are all working toward the same aim in science."[3]

The duality between competition and cooperation at CERN became clearer to me when, sometime later, I met Guido Tonelli, who heads the CMS group. "What's important is the concept of *fair competition*," Tonelli told me, "We must share all the information from our experiments—I would never imagine hiding our results from Fabiola, who runs the competing team at ATLAS. This is the only way to do science well. We need to compare our results for verification. And sharing information among competing teams is the only way to control our information. It's the best way to ensure that what we find in our experiments is true." Then he added, with a smile, "Besides, we are friends."[4]

What Tonelli told me had deeper implications, as I later learned. There have been rumors at CERN for decades that a certain team of researchers that once worked here had, in fact, withheld information and deceived its competitors, thus gaining an unfair advantage in the race to a scientific discovery. But this was all in the past, and as we move closer to finding the Higgs and supersymmetry and other discoveries in physics, the teams at CERN are working in an atmosphere of healthy competition and cooperation.

"We've managed to create unbelievable things here," Petagna said as we continued to discuss the miracle of CERN over coffee in the main building. "This consortium of the universities and institutions that sponsor the science projects here has brought physics to a world scale," he said, referring to the creation of CERN, its mission, and its multinational committee that decides on goals and means to achieve them and in many ways charts the future course of physics.

He added: "Another key point about CERN is the strong, direct link we have here between theory and experimentation. Theory decides what things to look for: the Higgs boson, supersymmetry, et cetera.

And then, the results of experiments tell the theorists what to work on. There is a continuous interplay between theory and experiment."[5]

Paolo then took me to meet the head of theory at CERN, Dr. Luis Alvarez-Gaume. We had a very pleasant discussion in his office on the top floor of the building. "My research is on the jets that appear above and below the massive black holes in the centers of galaxies," Alvarez-Gaume said when I asked him about his work. I thought that was a curious coincidence, given all the public worries about the possible creation of a black hole in the LHC. He laughed when I mentioned it; there's nothing like a little black hole to get physicists to relax and talk about the wonders of science. "Sure," said Luis, "we are also able to look for the signature of a micro black hole here at the LHC, but the machine was built to look for the Higgs, and so many other things.

"It's a Higgs factory," Alvarez-Gaume went on to say.[6] Over the following months I heard scientists describe the LHC as a *factory* of many different products—jets of quarks (stable quarks are the constituents of the protons and neutrons that live inside nuclei of matter; heavier, unstable quarks are created in accelerators and presumably existed in the very early universe); leptons, which are electrons or other elementary particles that are similar to them; taus, which are unstable leptons that are almost 3,500 times heavier than electrons; dark matter, which is composed of unknown particles scientists hope to identify; and neutrinos, which are extremely light, neutral leptons—depending on the researcher's taste or predilection. Scientists often view an accelerator as a factory, for once a particle is created, which requires a certain amount of energy, it will continue to be created—with a frequency that can be high or low, depending on the conditions—as long as the machine keeps running. So just what the LHC is a factory of, what kinds of particles it produces most often, is at this point an open question. But it became clear to me that every physicist thinks of the LHC as a factory of the particles he or she desires in order to pave the way to a big discovery.

I also learned that Alvarez-Gaume has interesting ideas about the structure of the universe. "There doesn't seem to be a design in the masses, the hierarchies," he said, alluding to one of the big problems in particle physics: the question of why the masses of particles vary without an apparent pattern. "Maybe the universe has no purpose at all," he said.[7]

I asked Alvarez-Gaume about the speed of the protons in the LHC. "If one of our protons from the LHC will race against a photon [a particle of light, traveling at the maximum speed possible in the universe] to the nearest star, Alpha Centauri," he answered, "the light ray, the photon, will beat our proton to that star by only 0.3 seconds!"[8] Alpha Centauri is 4.2 light-years away from us, which gives us a good idea about just how fast the LHC protons are moving. My friend Barton Zwiebach, a string theorist at MIT, has computed that a photon would beat one of the LHC protons by a mere *quarter of a millimeter* after completing one lap around the 16.5-mile track of the LHC.[9] The two analogies, by Zwiebach and by Alvarez-Gaume, are equivalent. And both demonstrate just how immensely fast these protons move inside the tube of the collider.

The LHC accelerates protons to a maximum speed that is very close to that of light, but many prior stages are needed in order to get the protons to this incredibly high speed.[10] The process is similar to that of using a rocket to send a spacecraft to Mars. The rocket has several stages, and each successive stage gives the spacecraft carried by the rocket an additional "kick" that increases its speed until it reaches escape velocity from our planet. The same is true of the protons accelerated at CERN.

In preparation for the LHC, the protons are first accelerated in a number of CERN's older accelerators, which are now being used as feeders for the gigantic LHC. First, the protons are created from atoms of hydrogen gas by stripping away electrons and leaving just the protons (which are positive hydrogen ions). The proton beams produced from the hydrogen gas are very intensive, with billions of protons in each batch, and yet every day of operation the LHC uses a mere 2 nanograms (two billionths of a gram) of hydrogen. To use a single gram of hydrogen, the LHC would have to operate continuously for more than a million years.

The protons are then given their initial boost of speed in the relatively small *linear* accelerator (rather than a circular one, such as the LHC), called Linac2. This accelerator gets the protons to 31.4 percent of the speed of light—that is, it gives them a velocity of 56,000 miles per second. This is a tremendous speed—faster than anything on Earth (other than light, or radio waves, which travel at the speed of light, or cosmic rays from outer space) but far slower than particles in most high-energy accelerators.

Once they reach their target velocity, the protons go from the linear accelerator into CERN's old circular particle accelerator, called the Proton Synchrotron (PS) Booster. Circular accelerators have the advantage that a particle can stay inside them for a long time, increasing its speed on every lap. In a linear accelerator, once the particle has reached the other end of the accelerator, that's it. The PS Booster brings the protons' speed up to 91.6 percent of the speed of light. Then a more powerful circular accelerator is needed for further boosts of speed. It should be mentioned that incremental increases in the velocity of the particles are increasingly expensive in terms of the energy they require. To get a proton from 31.4 percent of the speed of light to 91.6 percent of the speed of light (a difference of 60 percentage points) is much easier than getting it from 91.6 percent to 99.93 percent of the speed of light (a difference of only 8.3 percentage points), which is done by the Proton Synchrotron (PS). The reason that successive increases in speed require greater energies has to do with the special theory of relativity. As a particle moves faster and faster, effectively its mass grows—it seems to weigh more and more—and a heavier thing is harder to accelerate than a light one.

After the protons reach their target velocity of 99.93 percent of the speed of light inside the PS machine, they are funneled into the next-larger accelerator. This is the powerful Super Proton Synchrotron (SPS), which in 1983 revealed the existence of the W and Z bosons— particles that mediate the action of the weak nuclear force. CERN physicists Carlo Rubbia and Simon van der Meer received the 1984 Nobel Prize in Physics for their work leading to these discoveries. The SPS gives the protons a further kick of energy, accelerating them to 99.998 percent of the speed of light. This is equivalent to bringing the protons to an energy level of 450 GeV.

At last the protons are ready to enter their final destination: the LHC. The LHC itself can bring the protons to a maximum speed of 99.9999991 percent of the speed of light when it operates at its maximum energy of 14 TeV. It can give each proton a maximum energy of 7 TeV, thus as much as fifteen and a half times the total energy they entered it with, which is 450 GeV (one TeV is 1,000 GeV).

This acceleration by the LHC takes an immense amount of energy. By analogy, accelerating a 70,000-ton ship takes a large amount of fuel, while a car needs a small fraction of a gallon of gas to get it from 0 to 60 miles per hour. The acceleration of particles inside an accelerator follows exactly the same principle. To get a particle to a certain speed from rest takes some energy. But once the particle moves very fast, effectively it is no longer like a small car but rather like a heavy truck, so to get it to move still faster takes more energy. Then to get it to move even faster takes much more energy, for in our analogy its mass is now like that of a large ship.

To take a massive particle (that is, anything but light or a radio wave) and accelerate it to the speed of light itself is impossible because the particle's effective mass becomes infinite at the speed of light, so it would require an infinite amount of energy to get it to reach that speed. But to accelerate a proton to 99.9999991 percent of the speed of light—not 100 percent, but still close—can be achieved if you have at your disposal an energy source equivalent to the electric supply of an entire city the size of Geneva. And CERN does. This electrical energy is transformed by the laboratory into electromagnetic "pulling" energy through the work of the radio frequency devices that accelerate the particles; the nearly 10,000 giant superconducting electromagnets laid out throughout the 16.5 miles of the LHC tunnel then keep the protons on track and focused in a very tight beam over the many millions of laps they make inside the tunnel.

What the radio frequency devices inside the LHC cavity do is akin to what an adult does when pushing a child on a swing: Every time the child comes close, the adult pushes the swing, giving it more energy to increase the speed. Of course, adults stop pushing hard after a while so the child won't fall and get hurt, but the protons entering into the Large Hadron Collider continue to be pushed by radio devices every time they pass by. A perfectly synchronized pulse of energy keeps accelerating the protons faster and faster while the magnets bend, correct, and focus the beams, until the particles reach the maximum speed possible, which depends on the electrical energy available to the LHC, as well as engineering constraints on how much current the electromagnets can sustain.

The LHC, however, is often run at energies far below its maximum ability because the machine needs to be tested before it can be pushed to its limit. The superconducting magnets also need to be "trained" to run at high energies by gradually increasing their current input. The LHC is like a new car you just bought. If your car can be driven at 120 miles per hour, you won't want to strain it to reach that speed on the day you purchase it; first you'll want to get comfortable driving it and then break in the engine gently at lower speeds before you feel it can take the exertion of driving as fast as it can go. The LHC is a far more complex and sensitive machine than a car is, and unlike a car it is unique—there aren't thousands like it, so everything has to be learned from scratch.

The LHC is constructed of millions of times more working parts than an automobile, all of which need to work in perfect harmony for it to perform its task. This is why the machine needs a long preparation and testing period to get all its various components working as one. This is hard to achieve, since a low performance by one small element can derail the entire enterprise. As we've seen, a mere welding point—of which the LHC has many thousands—made the whole operation come to a dead stop when it failed.

One of the interesting things about the LHC and its construction is that no one really knows the properties, peculiarities, strengths, and weaknesses of this colossal creature. Scientists say it has a temperament, a unique responsiveness, even a spirit. And because it was put together in components by several different teams of scientists and engineers, its various parts have different personalities that need to be trained to work together. When I heard this description from people at CERN, I was reminded of a couple I knew in Alaska who had built a cement-and-steel sailboat. Both were experienced sailors, but when the boat's construction was finished, they had to test it slowly, at different wind speeds, and in a variety of weather conditions, before they could tell how it might perform in extreme conditions—such as crossing the stormy Pacific Ocean. They said they had to learn about the boat they had just finished building together. The testing of the LHC is somewhat similar, but of course on a much larger scale. The LHC has a personality too!

A sailboat can sink if pushed beyond its limits, and a car engine can burn if revved beyond what it can take. We know that the LHC can lose its magnets if the electric resistance goes up even slightly, but there are other operational problems that also might occur. For example, the proton beam might slip out of focus—that is, the protons might fail to stay within the very narrow range needed in order to crash them at precisely determined spots with the beam coming from the opposite direction. These protons fly around the 16.5-mile track an incredible 11,245 times every *second,* and they require twenty minutes to get them to maximum energy from the time they first enter the LHC tube.

In the collider, 100 billion protons are sent in each bunch, and there are 2,808 bunches in each proton beam. Bunches of protons are sent alternately into two parallel tubes inside the LHC: One beam of protons travels the 16.5-mile circuit in a clockwise direction, and the other travels counterclockwise. Once the protons have reached their maximum speed, they are directed to crash head-on into their counterparts coming from the opposite direction. The collisions are made at precisely determined times and locations along the circuit. Each bunch of 100 billion protons crashes inside one of three of the main detectors of the LHC, that is, at ATLAS, CMS, or LHCb. All the LHC collisions are highly localized at precise locations inside the detectors and release their energies within tiny volumes of space so that the release of energy is tightly controlled and the results of the collisions—cascades of new particles—can be accurately measured and studied.

The LHC has eight points along the tunnel for potential detectors. At four of these points there are actual detectors, and the other four are leftover cavities from the older, decommissioned accelerator called LEP—the Large Electron-Positron Collider—which gave its tunnel to the new LHC when it was dismantled in 2000. The working detectors are ATLAS (located at Point 1), CMS (Point 5), LHCb (Point 8), and ALICE (Point 2). But there are two additional, smaller projects at the LHC: One is called LHCf ("f" stands for "forward," denoting the forward location of the detector, near ATLAS); the other is called TOTEM (TOTal cross section, Elastic scattering and diffraction dissociation Measurement at the LHC) and is located near CMS. The LHCf experiment looks at collisions in an effort to use the results to model the behavior of cosmic rays, and TOTEM conducts size measurements of protons.

At LHCb, scientists use very sensitive devices known as Cherenkov counters to detect and measure light emitted when particles travel through matter at great speeds. LHCb looks for a special kind of decay process, the decay of particles called B-mesons. Mesons are intermediate-size particles—that is, particles with mass in between that of the electron and that of the 1,800-times-heavier proton; typically, they are a few hundred times heavier than an electron. Mesons are pairs of quarks and antiquarks, but B-mesons are particular mesons that contain the very heavy *bottom quark,* also called the beauty quark (quarks will be discussed in more detail later). B-mesons are the products of some of the collisions of the LHC protons, and they in turn decay into lighter particles. Scientists at CERN hope that by studying the decay modes of B-mesons they may learn something about why matter dominates over antimatter in the universe.

The ALICE collaboration is a group of scientists who use their specialized detector to study the conditions believed to have prevailed in the very early universe. In a sense, all of the Large Hadron Collider's work is like time travel to the very early universe. But what ALICE does is to perform a particular experiment. Ten percent of the time, usually for a short period after the machine has crashed protons for several months, the normal LHC proton beams are stopped and are replaced by beams of lead ions.

Why lead ions? Lead is among the heaviest elements on Earth, and once electrons are stripped from the lead atoms, the remaining ions are positively charged nuclei with a lot of mass. These are accelerated by the LHC to a lower velocity than the usual protons; because each of these ions now has a total weight of as much as 207 protons, it is harder to accelerate. But these ions still reach a speed that is respectably close to that of light. On the other hand, because the lead nuclei are so much heavier than mere protons, the collision energy of the lead nuclei is much higher. While for proton collisions the maximum LHC energy is 14 TeV, the lead nuclei crash into each other with a total of 1,150 TeV—an immense amount of energy when concentrated in a tiny volume of space inside the ALICE detector. When these heavy ions crash together, they create a plasma, a kind of fluid of particles some of which are electrically charged. The plasma is very hot and consists of quarks and gluons. This is the quark-gluon plasma, sometimes called "quark soup."

Quarks, which are the constituents of protons and neutrons,

are *confined* inside the protons and neutrons that contain them. Because the strong force holding them together is so ferociously powerful, quarks can never escape. Shortly after the Big Bang, however, the temperature in the nascent universe was so high that the quarks were immersed in the extremely hot plasma that permeated the universe before it cooled down enough for protons and neutrons to coalesce. The plasma also contained the force-carrying particles that bind the quarks together, appropriately named *gluons.* By re-creating the intensely hot quark-gluon plasma, whose temperature is in trillions of degrees—hundreds of thousands of times greater than that inside the center of the Sun—the ALICE collaboration hopes to learn about the behavior of the quarks and gluons, and hopefully discover how they settled to pro-

An aerial photograph showing the track of the LHC, and its detectors

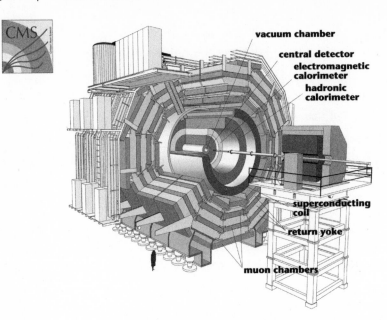

A schematic diagram of the CMS detector (with a person at the bottom to show scale)

duce the protons and neutrons that make up the nuclei of all the matter we see in the universe today. Similar research has been carried out in the United States, at the Brookhaven National Laboratory, where a quark-gluon plasma was reported to have been created briefly in early 2010.

The Compact Muon Solenoid (CMS), where more general research is conducted, is a compact detector in the sense that it packs a lot of electronic hardware in a confined space. The muon is one of the particles it detects—an unstable particle that resembles the electron but weighs more than two hundred times more—and a solenoid is a coil electromagnet. The magnet in the CMS detector creates an immensely powerful magnetic field of 4 tesla.[11] By comparison, the Earth's magnetic field, as measured on the surface, is a hundred thousand times weaker than that produced by the CMS detector.

ATLAS, the other general-purpose detector, is designed as a powerful toroidal (doughnut-shaped) superconducting magnet. Inside it are many thousands of detection components that provide very high reso-

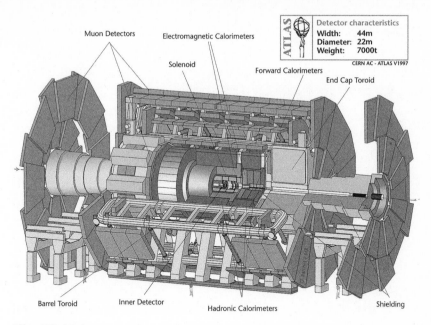

Muon Detectors   Electromagnetic Calorimeters

ATLAS

Detector characteristics
Width: 44m
Diameter: 22m
Weight: 7000t

CERN AC - ATLAS V1997

Solenoid

Forward Calorimeters

End Cap Toroid

Barrel Toroid   Inner Detector

Hadronic Calorimeters

Shielding

The ATLAS detector

lution for determining the tracks of particles created in proton collisions in the chamber at ATLAS's center. Like CMS, ATLAS has a
solenoid, the working part of a magnet, which turns electric power
into a strong magnetic field that bends the paths of the charged particles it detects. There are detectors at various levels inside ATLAS.
The *inner detector* charts the trajectories of charged particles that come
out of the collisions. Then there is a calorimeter, which measures the
energies of the particles produced. Even though they are heavier
than their cousins the electrons, muons are not stopped by the calorimeter and move farther out. There, they are measured by the muon
spectrometer that surrounds the calorimeter, which is very accurate
and gives excellent measurements of the momentum of the muons that
enter it (the work of the ATLAS detector is described in more detail in
Appendix A).

Both the ATLAS and the CMS detectors have been adjusted to
account for noise—cosmic rays and radioactivity from the ground and
air and other sources of stray signals—so that what they actually do

measure comes only from the proton collisions. After forty hours of testing with cosmic rays, the performance of the detectors was well understood. Then came the test with low-energy proton beams on September 7 and 10, 2008, and the results were excellent.

The CMS detector was tested continuously—or almost continuously—for 19 days. As CERN scientist André David showed a graph of the data of the test run to a lecture hall full of many of the world's leading particle physicists, he apologized for two flat areas in the display. "You see," he said, "we had to stop for a day. We had a VIP visit: The French prime minister and a German minister came to visit the LHC . . . a bit of a bummer for us. But, well, you take the opportunity to go home and visit your wife and make sure she's not too upset that you've been in the lab continuously for so many days."[12] Then, pointing to the next flattening of the graph of the data collected at CMS, he said: "And this occurred a few days later, when one of the magnets decided to quench and leak 6 tons of liquid helium."[13]

The big quench of September 19, 2008, was a blow to the many scientists who have worked at CERN for more than a decade with admirable dedication to their goal. Everything they had done for such an intense, long period of time was in preparation for proton collisions, and the quench not only caused millions of dollars' worth of damage and set the project back by over a year, it hurt their collective pride. They knew the world's eyes were on them, and their magical machine had let them down. They realized that they had to correct the structural flaws that had led to the accident and that they had to learn more about the capabilities of the collider so that such an event would never happen again.

Beyond the technicalities, politics had played a role here as well. The director general of CERN from 2004 to 2008 was the French executive Robert Aymar. He was reportedly a tough-minded man who pushed the operation at CERN to its limit. His ambition seems to have been to get the machine to operate at full capacity as soon as possible so that it might start producing results while he was still in office. But the LHC wasn't ready for that. "It has taken years to build the LHC," said Paolo Petagna. "It made no sense to rush." Now CERN is headed by the German national Rolf Heuer, and he seems to have a better understanding of the need for caution and a progressive start of the operation. "He is very well liked, seems to understand the machine and the people well, and likes to listen," Petagna told me.[14]

This is how Paolo Petagna explained the occasion of the restart of the LHC in November 2009:

> You know, it's a bit like a high jumper at the Olympics: He goes in relaxed when he starts 20 centimeters under his seasonal best; but when it comes to jumping for the gold medal and the world record . . . you can bet that he takes all his time and takes precautions before running! In particular, if the last time he tried it, he rushed too much and crashed . . . The restart of the machine [in November 2009] has been brilliant and really above all our expectations—but now we had better triple-check every detail and do everything to avoid any bad moves before moving on to the real unknown![15]

I met Fabiola Gianotti, the spokesperson of ATLAS, the friendly competitor of CMS, in her office at CERN headquarters with its views of the grassy area below, the other buildings of CERN, and the ubiquitous high-power lines. It was a comfortable office on the fourth floor of the building of the ATLAS and CMS groups, where the physicists have their offices, meet in halls to discuss their ideas and results, and perform their analyses. Gianotti received wide media coverage in her native Italy when she took her position at the helm of the ATLAS team.

The first thing Gianotti told me when I asked her what she thought the LHC will find was: "Nature is often more elegant and more intelligent than human beings."[16] She then explained what she meant. People build up theories and expectations about what nature will do—about what they should find in their experiments. But nature does what it wants, so the experiments will often bring big surprises. "This is how we learn about the beauty of nature and its intelligence in creating unexpectedly elegant and meaningful results beyond anyone's expectations. The most exciting thing about research is to have an open mind," she said. "Columbus left looking for India and instead found America."[17]

It is this penetrating vision and fresh approach to science that has endeared Fabiola to millions of people in Italy, where she is a folk hero whose scientific career path is carefully documented by the press. Fabiola Gianotti received her doctorate in physics from the University of Milan in 1988 and afterward took a position at CERN, where she worked on a crucial instrument used in the ATLAS detector. She is a lover of music and dance and still spends her free time playing music. "There will be no black hole produced here," she assured me when I asked her the question she gets asked most often by the press. "Cosmic rays have been interacting in outer space for billions of years, with much more energy than what we will produce in the LHC."[18]

Gianotti also wanted me to know that the value of CERN goes far beyond pure science and the acquisition of knowledge about the universe. "We build complicated technology here, and then our technology is used elsewhere, in industry all over the world. We have a partnership with industry so that our research can be used to better society and improve life. What we have with the LHC project is an international collaboration that for the first time in history is truly on a global scale."[19]

CERN brings many doctoral students and young researchers from around the world to work here. This gives the scientists a chance to work in a productive global research environment. "It is very enriching for our students, who come here from all over the world, to 'grow up' professionally in a place with a cooperative and tolerant attitude," Gianotti said, "and this cooperation helps boost the technological level of the countries the students come from. All these aspects make this place very special."[20]

My first visit to CERN was leaving me speechless. I had never imagined that a place like this might exist—one where ten thousand scientists from around the world work together in great enthusiasm and harmony to pursue the ultimate knowledge about our origins and the nature of reality. This environment is so exceptional—not only for science, but also as a general organization of human beings—that it has attracted close attention from anthropologists, sociologists, and historians, all of them trying to understand how CERN works in terms of its people and their interactions and decision making.

A social analysis of the human element at CERN—the concentration of thousands of scientists working together in one place, rarely leaving it, and socializing and interacting among themselves—has recently been published in the scientific journal *Nature,* titled "The Large Human Collider."[21] The article pointed out that never before in history have so many scientists congregated together and worked as a group, or rather as multiple, mutually interrelated groups, all striving toward one aim. Because every scientist is an expert in a field, the usual top-down management structure of business or government or military organizations does not work here, and CERN had to invent its own way of working—a much more harmonious and cooperative one. "The industrial model cannot work. One human simply cannot make technical decisions on such a large scale," according to an anthropologist who has studied the workings of CERN.[22]

The functioning of the scientists here has had strong implications for society that go far beyond the actual science pursued in this enchanted place. Caught up in its contagious enthusiasm, I wanted to know more about how the Large Hadron Collider was conceived and constructed. And I wanted to learn more about the history of the CERN organization that has made the LHC possible.

Chapter 4

Building the Greatest
Machine in History

Paolo Petagna took me to the CERN security office to arrange for an ID that would allow me to enter the cavity of the CMS detector, 300 feet underground at Point 5 of the LHC. I had to fill out some forms and was then given an identity card to allow me through into the belly of the detector. We drove through the Swiss and French countryside to Point 5 and entered an elevator that would take us deep down into the detector. There were radiation signs and warnings about security in some places.

"Doesn't really look quite like *Angels and Demons*," I remarked. Paolo laughed. "Well, you should know that Dan Brown did visit here, but there were no eye scanners here at that time. I don't know whether the book and movie gave them the idea to install these security checks or whether these had already been planned, but the reason for this is very different from what was presented so dramatically in that story," he said. He explained that if someone is in the LHC tunnel and it needs to close, it's important for the computer to know about it so that the person can be found quickly and brought back to the surface. It's a matter of personal safety—no one should be accidentally left inside a locked tunnel three hundred feet belowground. We reached the floor of the detector and came out of the elevator. What I saw next stunned me.

My friend Barton Zwiebach had described the LHC to me as "the most unbelievable machine ever built by humanity. It ranks with the cathedrals of Europe."[1] So now, visiting CERN with Paolo as my guide, seeing the sophisticated instrumentation, the precision electronics, and the sheer mass and complexity of the whole construction, I thought about what Barton had told me when I had lunch with him at MIT just before my trip to Geneva. I finally began to really understand what he meant—and I could not agree with him more.

Standing deep underground, looking up from the floor of the tunnel at the massive detector that rises to the height of a five-story building—a structure that had taken so much effort and talent to construct—certainly reminded me of standing on the Île de la Cité in the heart of Paris, admiring the Notre Dame Cathedral rising above the square. In both cases, you marvel at the sheer immensity of the construction and at the human ingenuity, drive, ambition, and aesthetic taste that went into building it: here, in a twenty-first-century project of science, and there, as the apex of medieval art and architecture in the twelfth century.

The exact total length of the LHC is 26,659 meters (26.659 kilometers, or 16.565 miles). I was surprised to learn that far from being the world's largest tunnel, this one ranks twenty-first. The longest tunnel on Earth is in the United States: The Delaware Aqueduct, in the state of New York, which is 85.1 miles long, drilled in solid rock, transports water. It is followed by the Päijänne Water Tunnel in southern Finland, which is 74.6 miles long. And the Orange–Fish River Tunnel in South Africa, which also supplies water, is 51.4 miles long. In Switzerland itself, the Lötschberg Base Tunnel, in the Bernese Alps, completed in 2007, is the longest railway tunnel, at 21.5 miles. And even the Madrid metro system, the Berlin U-Bahn, and the Montreal metro in Canada each has a single tunnel longer than the LHC. There are other railroad and subway tunnels in Russia, China, Spain, and London that are longer than the tunnel of the LHC, but the LHC is the world's largest scientific instrument and the largest machine ever built.

Inside the original tunnel were to be placed—through incredible logistical and engineering feats—the LHC's almost 10,000 large electromagnets, cooled by 10,080 tons of liquid helium to the superlow temperature of 1.9 degrees Kelvin (–456.25 degrees Fahrenheit), making the LHC the coldest place in the universe. Of course we haven't visited every location in the universe to confirm this, but this assertion does have some logical support. Our universe was created in the intense heat that followed the Big Bang some 13.7 billion years ago and has been cooling down ever since. The present temperature of the universe, which is 2.73 degrees Kelvin (-454.7 degrees Fahrenheit), represents the "embers of the Big Bang"—it is the temperature that remains at the present epoch as the remnant of the continuously cooling heat of the Big Bang. But inside the LHC, the superconducting magnets require a colder temperature than that of outer space in order to function properly, and they are therefore constantly being actively cooled. If another civilization somewhere is using energy to sustain temperatures lower than those of outer space, then the LHC may not be the coldest place in the universe, but otherwise this is it.

Inside the proton tubes in the tunnel of the LHC is a close-to-perfect vacuum—a pressure ten times lower than the pressure on the surface of the Moon. This vacuum is achieved by pumps that remove most of the air from the tubes in which the protons are accelerated so that they will not hit air molecules—only the protons accelerated in the opposite direction, when they meet inside the chamber of one of the detectors.

The two main parts of the CMS detector had been separated so that workers could make fine adjustments to the electronics inside them. We climbed up a scaffold placed between the two massive components of the detector. Here, Paolo showed me the actual cavity inside of which the protons crash against each other when the device is closed and the machine is running (and when everyone is safely aboveground and the tunnel is locked). He described the immensely powerful magnetic field that surrounds the crashing protons when the LHC is operational.

Paolo told me that a German colleague of his once insisted on staying inside the CMS cavity when everyone was told to leave as the magnetic field was turned on to 4 tesla in a test of the superconducting magnet. The physicist later described that his whole body ached from the intense magnetism that surrounded him, and his head felt very strange. "But if you stay put and try not to move at all, it is bearable," he said.[2]

ATLAS is bigger than CMS, and it sits in two shafts. It is 151 feet long and 82 feet high—it's as high as a seven-story building. To dig out its cavity, 300,000 tons of rock had to be excavated—weighing as much as the Empire State Building. The entire detector is made of relatively lightweight materials (as compared with rocks) and still weighs 7,000 tons. There are 100 million active parts inside the ATLAS detector—all very densely packed together. And all the sensitive electronic components and other parts of the device are internally connected through almost 2,000 miles of electric cables.

"ATLAS was built like a ship in a bottle—piece by piece," said Dr. Peter Jenni, the Swiss particle physicist who had directed the construction of the detector and headed the ATLAS group until early 2009.[3] Very slowly, each large component of the detector was separately lowered down into the cavity 300 feet belowground, there to be put together and connected with the other parts. ATLAS is a huge magnet shaped like a torus with an eightfold symmetry; it is a giant doughnut, but made of eight coils symmetrically placed like spokes around the central hollow in which the protons crash. The eight connected magnetic coils work together to create the magnetic field needed to bend the paths of particles inside it. On October 26, 2004, a large crane lowered down into the tunnel the first of the eight large coils of the ATLAS detector, each of them 82 feet long and weighing one ton.

Putting together the ATLAS detector was a very tricky engineering task since each coil had to be twisted at an almost impossibly sharp angle so it could fit into the 62-feet-diameter vertical shaft that descended to the depth of the LHC cavity. Once they were finally belowground, the components were put together and attached to a horizontal platform. The level of precision with which the ATLAS detector was constructed is better than one-hundredth of a millimeter.[4]

On February 28, 2007, the moment of truth in the construction of the CMS detector had arrived. At six a.m., the heaviest piece of the detector began its final journey into the depths of the tunnel that once housed the LEP and was now to become the Large Hadron Collider. A massive gantry crane built by the Swiss company Vorspann System Losinger Group slowly entered the specially designed building sitting over the CMS cavity. The building had been constructed with the purpose of housing the gantry crane in this particular task and needed a variance from French building codes to allow it to stand as high as it does. Once the crane was inside the building, it began to lower its payload into the cavity deep underground. The preassembled central piece of the detector, containing the giant electromagnet, weighs 1,920 tons—as much as five Boeing 747 Jumbo Jets.

The entire detector weighs 12,500 tons and is the heaviest instrument ever built. Smaller than ATLAS, it is only 69 feet long and its diameter is 49 feet. Placing the central part of the detector, a huge piece of steel, copper, and niobium-titanium filled with extremely intricate and sensitive electronics, in its exact location underground proceeded at a snail's pace—30 feet an hour. It took more than a full working day—ten hours of nerve-racking precision work—to complete. But the CMS team successfully placed the magnet within a few millimeters of its intended position, so the required corrections were minimal. Putting the CMS in place was the critical moment in the entire construction project of the LHC—the stage that took the longest and that delayed many that were to follow.

At both the CMS and ATLAS, there are three hundred racks of electronics that measure, digitize, and transmit the signals from the particles that impact the various parts of the detectors. And above-ground, some 300 feet above the tunnel of the LHC (throughout its circuit, the depth of the tunnel varies from 160 feet to 575 feet), there is an additional set of one hundred racks of computer equipment.

A computing grid performs all the heavy computations, and the whole experiment generates 10 to 15 petabytes (a petabyte is a quadrillion, or 10^{15}, bytes) of information each year. This huge amount of data is analyzed on the Worldwide LHC Computing Grid, a two-tier system of more than 100 computing centers around the world. On-site, the analysis consists of scientists studying advanced three-dimensional multicolor graphics on large screens. The screens display images of the various levels of the detectors underground, showing the paths of the particles that result from the proton collisions. The computer analyzes the direction of each particle's track through the various parts of the detector. The particle's path is made to curve by the intense magnetic field applied to it by the detector, and this curvature is measured and aids in identifying each particle. Subsequent analysis leads to a determination of each particle's mass, charge, and energy.

Typically, the scientists see magnificent displays of cascades of particles as each decays into others, all of them making their own colored tracks on the screen. It looks like a fireworks display. The reason is

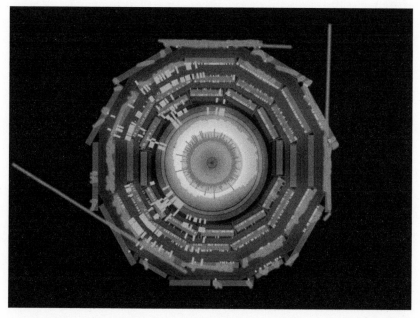

An early "splash" of particles in the CMS detector in 2008 obtained from collisions on a metal target

that most of the particles that emerge from the proton collisions are unstable and quickly decay into lighter particles, which, in turn, decay into lighter ones still.

A complete analysis of all the particles resulting from a collision, with all their decay products, leads to the potential identification of "new" particles—those that have never been seen before. But these are rare, and scientists expect that only a few out of the many collisions in the detectors of the machine may lead to a potential new discovery. At full capacity, the LHC produces many trillions of collisions every day, so the charge of the vast worldwide computing system used by CERN is to sift through the results of the immense number of collisions produced here in search of the few that are important—those that can lead to new discoveries about nature. It's a supermodern, computerized search for the proverbial needle in a haystack.

You might well wonder how two detectors, ATLAS and CMS, which are aimed at searching for exactly the same phenomena, could be designed so differently from each other—one detector being very heavy and relatively small, the other twice as long and more than one and a half times higher with a little more than half the weight; one an eight-fold toroid, the other looking like a polygonal cylinder; one operating at a magnetic field intensity of 1 and 2 tesla, the other at a much more intense magnetic field of 4 tesla. There is an interesting reason for these differences. Both of these superconducting detectors were designed in the 1990s by their respective teams of scientists. At that time, finding the Higgs boson was considered the main purpose for the construction of the LHC, and both teams were charged with designing state-of-the-art particle detectors that could achieve this major aim of the collider. The most likely way that a Higgs could be detected, it was thought, was through its decay into two Z bosons, which in turn would disintegrate into four muons.[5]

Both teams, therefore, searched for the best design that would allow them to detect a cascade of particles decaying into four muons, and their benchmark was to maximize the ability of their respective instruments to detect at once four muons traveling within the detector with measurement errors no worse than 10 percent. The mathematical formula for calculating the ability of a detector to find and measure particles with high accuracy is magnetic field strength times the square of the distance a particle travels inside the detector.[6] The two teams realized that the equation had two distinct solutions within the constraints they were given: Either choose a compact instrument, which offers a relatively short path for the muons inside the detector, but with a very powerful magnetic field to bend this path; or choose a much bigger instrument, which would allow the observation of a muon as it traveled a longer distance inside the detector, in which case a much smaller magnetic field intensity would suffice.

The CMS team chose the first solution for its design, and the ATLAS team chose the second. At CMS, the muons travel 3 meters (about 10 feet) while an intense field of 4 tesla acts on them. The detection power for muons in CMS is thus $4 \times 3^2 = 36$. At ATLAS, while in the inner chamber the magnetic field is 2 tesla, in the muon chamber the field is only 1 tesla; but this chamber is twice as large (in distance traveled) as the muon chamber of CMS, and the muons traverse 6 meters (about 20 feet) inside it. So the detection power for muons in ATLAS is $1 \times 6^2 = 36$. The two competing projects have identical detection power! Time will tell which design is better, if there is a better one—or which of the two teams has better instrumentation or computing power, or more astute observation skills, or luck.

A particle's electric charge is determined by the *direction* of the curvature of the particle's track in the magnetic field of the detector: Tracks of particles with negative charge curve in the opposite way from those of particles with a positive charge. A particle's momentum can be measured by the degree of the bending of the curved path. The track of a fast-moving particle curves less than that of a slow particle because a fast particle spends less time inside the magnetic field that exerts its bending force on it. A neutral particle's path is not bent at all because it is the electric charge that makes a particle swerve inside an electromagnetic field.

As we have seen, for purposes of identifying particles, a field of 1 and 2 tesla is enough for ATLAS, and the smaller-size CMS uses a field of 4 tesla. But the LHC uses superconducting magnets for other purposes: The 1,232 dipole magnets bend the two proton beams traveling in opposite directions inside the two parallel tubes along the 16.5-mile circumference of the LHC. Each of these magnets is 47 feet long, and they require an even stronger magnetic field in order to do their difficult job of making massive particles travel in precisely determined curved paths at speeds very close to that of light. The superconducting dipole magnets require a magnetic field of 8.33 tesla, more than twice as strong as that of the CMS detector, and 200,000 times the strength of Earth's magnetic field. To achieve such a strong magnetic field, at top power, the dipole magnets need a current of 11,700 amperes.

These magnets have dual apertures inside them, one for each of the two opposing proton beams; this way, the same magnet controls the protons in both tubes—the ones traveling in a clockwise direction inside the tunnel and those moving counterclockwise. Recall that at the detectors, the two opposing beams come together and the protons are smashed against each other, so there is only one channel inside each of the detecting electromagnets. The particles are shunted to a detector—be it ATLAS, CMS, ALICE, or LHCb—at specific times after they've been accelerated and their tracks corrected and focused. The two smaller detectors, LHCf and TOTEM, depend on ATLAS and CMS for their particle collisions.

The magnetic field inside the superconducting magnets is created by niobium-titanium coils, which become superconducting when cooled by the liquid helium encircling each magnet. The total length of all the cables inside the LHC's dipole magnets is 4,700 miles, and they weigh 1,200 tons. Each cable is woven of 36 strands, and each strand is made of 6,400 tiny filaments of the niobium-titanium superconducting alloy, each of them only 0.007 millimeter thick—ten times thinner than a human hair. The total length of all these filaments is 1 billion miles. This is more than ten times the distance from Earth to the Sun

(10 AU, in astronomical distance units)![7] These statistics should give you an idea about the scope of the technological challenges that were faced by the people of CERN when they built the Large Hadron Collider. But the LHC project was only the latest and grandest in a long line of physical science research endeavors that began after the end of World War II.

––––––

After my talk with the ATLAS spokesperson, Fabiola Gianotti, Paolo Petagna took me to meet Nobel Laureate Jack Steinberger, who has been associated with CERN for decades. Steinberger told me that CERN was founded as a direct response to the Manhattan Project, in which the United States produced the first atomic bombs. The European nations wanted to launch a joint nuclear research project whose aim would be the peaceful exploration of the atom and its constituents. "The existence of CERN," Steinberger told me, "is therefore a direct counterreaction to the atomic bomb developed in America."[8]

Indeed, CERN has a rich history. Four years after Hiroshima and the end of World War II, an international cultural conference took place in Lausanne, Switzerland, just across the lake from the city of Geneva. Here, Prince Louis de Broglie, the quantum theory pioneer who proposed that all small particles behave like waves, first raised the possibility of creating a European science laboratory. This idea was enthusiastically embraced by the many participants in the meeting.

The following year, at the Fifth General Conference of UNESCO (the United Nations Educational, Scientific, and Cultural Organization), held in Florence, Italy, the American physicist and Nobel Prize winner Isidor Rabi put forward a resolution to establish exactly such a European-based physics laboratory, one that would be dedicated "to increase and make more fruitful the international collaboration of scientists."[9]

The reason that many scientists felt there was a strong need for this undertaking was that European physicists were then becoming more aware of the full extent of the brain drain suffered by their continent before and during the Second World War—due mainly to Hitler's purges of Jews from academia, but also to many other scientists fleeing wartime conditions in their ravaged countries. The participants in the UNESCO meeting understood that the effects of the wartime brain drain could continue to set back European science unless something drastic was done to reverse the trend and entice scientists to Europe.

Two more UNESCO conferences followed, in which the details of the proposal were fleshed out. Already in these early meetings, the surprising ability to reach a consensus among people of differing points of view and varied backgrounds, which is now characteristic of CERN, began to emerge. On February 15, 1952, at the UNESCO meeting held in Amsterdam, the representatives of eleven European nations signed an agreement setting up the Conseil Européen pour la Recherche Nucléaire (CERN). This initial agreement also established the Council of Representatives of European States for Planning an International Laboratory and Organizing Other Forms of Co-operation in Nuclear Research and gave it the charge of writing up a convention for the organization, as well as choosing a site and arranging for procurement of equipment and other necessities.

After lengthy discussions, the delegates decided that the new research body should be based at a location spanning the border region between Switzerland and France, two countries that offered space for the facility, and headquartered just outside the city of Geneva. The convention of the organization, whose official name was changed somewhat to the European Organization for Nuclear Research, but whose acronym CERN was retained, was formally ratified by its member states on September 29, 1954.

The original twelve founding nations sponsoring this international organization were Germany (Federal Republic at that time), Belgium, Denmark, France, Greece, Italy, Norway, Sweden, the Netherlands, the

United Kingdom, Switzerland, and Yugoslavia. Eventually, Yugoslavia left for financial reasons in 1961, but Spain, Portugal, Austria, Finland, Poland, the Czech and Slovak republics, Hungary, and Bulgaria joined by 1999, bringing the number of member states to twenty.

The purpose of CERN is to study particles through the use of accelerators. The idea of a particle accelerator goes back to late in the nineteenth century, when the British physicist J. J. Thomson discovered the electron—a negatively charged particle in the atom. This happened in 1898, when Thomson noticed that a ray he produced in a cathode ray tube was deflected in a magnetic field. He deduced from the experiment that the particles in the ray were negatively charged. These were the first electrons in history whose tracks were actually seen.

A cathode ray tube is what creates the picture on the screen in the old, bulky television sets (today's flat screens work differently). Electric potential is created inside a vacuum tube—a glass tube from which most of the air has been removed. The flat, picture end of the television tube is coated with material that emits light when hit by electrons. The electrons are accelerated in the electric potential inside the tube, thus creating a picture when the stream of electrons is made to sweep across the screen according to instructions that come through the UHF or VHF radio transmission, or mostly through cable nowadays, to the television set. The information in the signals is implemented by an electromagnetic field in the television set that rapidly deflects the ray of electrons moving toward the screen to form the desired picture.

This is the same basic idea behind the technology of a modern particle accelerator. Charged particles, electrons, positrons (the positron is the antiparticle of the electron), or protons—as in the case of the LHC—are obtained from matter and are accelerated and controlled through the use of electromagnetic fields. In the case of the LHC, the tube is circular and has a 16.5-mile circumference. Every time the particles pass by a magnet, its electromagnetic force bends their path. The radio frequency devices accelerate them. The protons of the LHC go through many millions of such cycles through the tube. But instead of crashing onto a fluorescent screen as in a television set or in Thomson's experimental apparatus that enabled him to discover the electron, they crash into other protons.

The superconducting magnets are needed to bend the paths of the protons because the LHC is circular. Linear accelerators need no bending magnets, although they do use magnets for focusing the beam and keeping it on track. As mentioned earlier, in a linear accelerator the particles travel through the straight tube only once and then crash. In a circular one, the particles can go through the circuit many times before they are made to crash. The disadvantage is that much energy is needed to keep bending the particles' track—the particles naturally "want" to travel in straight lines, so to keep them moving in a circular way requires high amounts of electromagnetic energy.

The idea of Thomson's experiment that revealed the existence of the electron also gives us a unit of measurement for energy that is understandable in terms of the experiment. The electrons that Thomson discovered are accelerated inside the vacuum tube through the use of an electromagnetic field. We define *1 electron volt* (eV) as the amount of energy that a single electron gains when it is accelerated in an electromagnetic field that has an electric potential of *1 volt*. The electric potential arises from the difference in electric charge between two points, expressed in the units of voltage we know from everyday life. In a 1.5-volt battery, for example, the potential difference between the plus and minus ends is 1.5 volts. The electricity in your house is at 110 volts. This comparison should give you some idea about the scale of electric potential: The voltage in the outlets in your home is seventy-three times more than that of a small battery. This is why you can touch both ends of a 1.5-volt (AA or AAA) battery but you wouldn't dare touch an exposed electric wire in your house—the electricity could kill you.

Imagine electrons flying inside Thomson's cathode ray tube. Suppose you attach your little 1.5-volt battery to his apparatus. How much energy do you think you would be adding to each of his electrons, as evidenced through their increased speed inside the tube? Since we have an intuitive feel that a 1.5-volt battery is not very powerful (you can touch both sides of the battery with your fingers without even feeling the electricity), clearly you will not have added much energy to the electrons, so we see that 1 electron volt (1 eV) is a very small amount of energy.

But now imagine that you use 1,000 volts. This is a strong voltage—you wouldn't dare touch a wire with such a voltage—and it imparts an amount of energy of 1,000 electron volts, or 1 kiloelectron volt (1 KeV) to the electrons inside the tube. Such a high voltage accelerates the electrons: It gives them measurable additional energy. Now imagine giving these electrons a *million* electron volts—that is, accelerating them through a potential difference of a million volts, which is a very high level.

A million electron volts, the amount of energy in electrons that have been accelerated through a potential difference of a million volts, is called a megaelectron volt (MeV); a gigaelectron volt (GeV) is the energy gained by an electron accelerated through a potential of a *billion* volts; a teraelectron volt (TeV) is the energy gained by an electron accelerated through a potential of a *trillion* volts; and a petaelectron volt (PeV) is the energy gained by an electron accelerated in a field with an electric potential difference of a *quadrillion* volts. We have already encountered three of these energy levels: The LHC's maximum energy is 7 TeV for each beam of protons, or 14 TeV in total when two opposing beams crash together. We've also seen that protons that come out of the SPS have an energy of 450 GeV; and recall that the energy equivalent of the mass of an electron at rest is 0.511 MeV. We are still far from being able to produce energy in the PeV range. Scientists today speak mostly about the TeV scale—the scale of energy produced by the LHC and also by the Tevatron accelerator at Fermilab, which can generate a total energy level of just under 2 TeV.

In 1932, the British researchers John D. Cockcroft and Ernest Walton built the world's first particle accelerator, a machine that accelerates particles to high speeds and smashes them. They wanted to "see" what was inside the particles. What was the nucleus made of? Did it have any component parts, and if so, what were these subatomic particles like?

There are two kinds of particle collisions: elastic and inelastic. In an elastic collision, both particles recoil from each other—think of two billiard balls hitting each other and going off in different directions: Nothing really happens to the balls themselves, and only their motions change. But in inelastic collisions, the balls break. This is what an accelerator does—it breaks particles of matter. In particle physics, an inelastic collision can be studied in terms of the energies produced, and as mentioned earlier, in very powerful collisions the particles themselves become energy; then in turn the total energy—from the mass of the crashing particles and from their speed—is transformed into other particles.

Cockcroft and Walton used an 8-foot-long vacuum tube and a power source of 800 kilovolts (800,000 volts) to accelerate protons to their target. They used a lithium target placed inside their accelerator, and they found that the protons completely disintegrated the lithium atoms, breaking them down into alpha particles (particles consisting of two protons and two neutrons). This was the first time that scientists were able to smash an atom—to break it into smaller parts using an accelerator (and this is why accelerators are sometimes called "atom smashers"). This was a major discovery, because it showed how the nucleus was put together: It was seen to have component parts that could be forced to emerge once enough energy was used to smash the nucleus. Because of the interplay between mass and energy in these reactions, the experiment also constituted a major proof of Einstein's famous formula $E = mc^2$.

The success enjoyed by Cockcroft and Walton provided the impetus for building more particle accelerators. The Russians and the Germans soon built their own, hoping to make further discoveries about the nature of the nucleus and how it was put together; they experimented with blowing up other kinds of nuclei. Through these experiments physicists were hoping to discover smaller and smaller particles that result from such collisions, to see what the tiniest building blocks of matter might be. The next major discoveries took place in California.

Ernest Lawrence, an American physicist from the Midwest, took a position at Berkeley after getting his PhD in physics from Yale in 1925.

He understood very little German but tried to read a paper in an ob-
scure German publication about the design of accelerators. In fact, he
deciphered the diagram in the paper without understanding the words,
which led him to the idea that a *circular* accelerator might work better
than a straight-line one. He then designed a 4.5-inch-diameter machine, and in 1931 built an 11-inch-diameter "cyclotron"—a circular
particle accelerator. In his cyclotron he was able to accelerate protons to
energies of a million electron volts (1 MeV), a major achievement. In
later years, Lawrence supervised the construction of bigger cyclotrons
at Berkeley, and through the research conducted in them certain radioactive elements were produced—for example, carbon 14. But more
powerful accelerators were required to release subatomic particles.

Research on accelerator design progressed over the decades as
scientists continued building both kinds of accelerators: circular ones
and those constructed in a straight line. In 1957, CERN launched its
600-megaelectron-volt (600 MeV) synchrocyclotron (a cyclotron that
uses a variable radio frequency device), which rivaled similar particle
accelerators developed in the United States. A major achievement of
the new machine was an experimental determination that a *pion* (the
first-discovered meson, a particle with mass between that of an electron
and that of a proton and composed of a quark and an antiquark) decayed directly into an electron and a neutrino.

In 1959, CERN launched the 28-GeV (28 billion electron volts)
Proton Synchrotron (PS), which reigned for some time as the most
powerful particle accelerator in the world. That same year, the first
CERN experiments with neutrino beams were inaugurated. The
elusive neutrino is produced in certain particle collisions, and when
enough neutrinos are made—in beams containing many trillions of
them—some neutrinos can be detected, even though the detection rate
for a single neutrino is extremely low since these neutral, tiny particles
are notoriously averse to interacting with other matter.

At Stanford University, the famous Stanford Linear Accelerator
Center (SLAC) was built in 1966, and through its operation new par-

ticles were found, among them a very heavy cousin of the electron called the tau particle, discovered the following decade by a team headed by Martin Perl. Heavy quarks were discovered at the Tevatron in Fermilab, a circular accelerator built in the 1980s. Two new kinds of neutrinos are also among the most important discoveries made through work in particle accelerators over the decades.

Because high-power accelerators are extremely costly to design, build, and operate, European nations found it necessary to pool resources in order to meet the costs. They had to decide on a budget scheme that would divide the cost of the operation proportionally among all the sponsoring nations, based on the relative sizes of their economies. No single European nation could have come up with the needed cash on its own as the United States has been able to do with its accelerators: the cyclotron at Berkeley, SLAC at Stanford, Fermilab's Tevatron near Chicago, and the accelerator at the Brookhaven National Laboratory on Long Island. These machines are run by universities but funded with federal money.

In 1993, the United States Congress discontinued funding for the planned Superconducting Super Collider (SSC) project in Texas, which was to produce energies higher than those available through the LHC. This move left the European CERN as today's global leader in particle research. It should be noted, however, that many of the thousands of physicists who work at CERN, at least part-time while visiting from their home institutions, are American. The United States, along with the Russian Federation, India, Japan, Israel, Turkey, the European Commission, and UNESCO, has "observer" status at CERN, and this allows many U.S.-based physicists access to the laboratory through their home institutions.

The science pursued at CERN is administered independently of the technology and the logistics. So while the United States is neither part of Europe nor a member state of this European concern, the scientific collaborations that are based at CERN do include many American scientists as well as those from other non-European nations. The CMS collaboration, for example, has many participating American institutions, such as MIT, the University of California, and others. In fact, about a third of all the scientists involved in research projects of the CMS collaboration belong to U.S.-based institutions.

CERN has built a number of accelerators of increasing energies over the decades, and in September 1989 it inaugurated the Large Electron-Positron Collider (LEP). The tunnel that now houses the LHC was originally excavated for the LEP. French president François Mitterand and Swiss president Pierre Aubert attended the launching ceremony of LEP. Unlike tunnels used for transportation, this cavity had to be constructed to exacting measurement specifications. When the tunnel's digging was finished, the two excavating teams working toward each other found that the two ends of the tunnel met nearly perfectly: The discrepancy was less than 1 centimeter![10]

When particle beams were first accelerated in the LEP, scientists noticed that at certain times of the day, the particles missed their rendezvous with the particles flying in the opposite direction. The physicists scratched their heads and the puzzle was studied by a number of teams. For a while, no one could figure out what was going on. Then the mystery was solved: The culprit was the tides! The gravitational pull of the Moon, which gives us the daily variations in the levels of the oceans, the tides, also affects "solid" ground. The ground near Geneva moves enough due to the tidal variations so that over a 16.5-mile distance in the tunnel the particles can be diverted and miss their mark. The problem was corrected using modern engineering methods to adjust the position of all the points around the circuit so that the tidal fluctuations are nullified by compensating countermotions. Another problem that was noticed was that the electricity level varied somewhat in the area when the high-speed French TGV train (Train à Grande Vitesse), which uses much electricity, approached Geneva. This variation too had to be corrected by CERN engineers.

In August 1989, LEP began operation, and within two months it led to important measurements of the Z boson. By December 1991, the delegates to the CERN Council unanimously decided that the organization should build the *next*-generation particle accelerator: the Large Hadron Collider. Before it was decommissioned in 2000, the LEP produced tantalizing results that some physicists interpreted as "hints of a Higgs," but there were just not enough data to confirm the existence of this particle. Scientists petitioned the organization to allow the LEP to operate a few months into 2001, hoping that more time might allow them to obtain conclusive data. But the LHC project was considered such a high priority that CERN's management turned down the request, and the LEP was dismantled to make room for the new collider.

In the mid-1990s, both the United States and Japan made generous financial contributions to CERN and earned the right to become observers at the CERN Council. In 1998, the first of the many 47-feet-long magnets that bend the proton beams of the LHC arrived at CERN. The first components of the ATLAS detector had to be brought into the cavity while the LEP was still operational, which presented great challenges in putting them in place.

The LEP took fourteen months to dismantle: 40,000 tons of material had to be lifted to ground level from deep below the surface. The United States shipped to CERN twenty special superconducting magnets; other magnets came from Russia and from other countries. Within a few years, all the components of the LHC, including the detectors, were in place, and the greatest machine on Earth was ready to begin its mission.

Chapter 5

LHCb and the Mystery of the Missing Antimatter

The LHC smashes together tiny particles given to the strange effects of quantum theory. But one of the major quests of CERN and the LHC is a deeper understanding of a phenomenon derived from quantum mechanics combined with Einstein's special theory of relativity: the existence of *antimatter*. We know today that all matter particles have bizarre mirror-image twins. The electron has a twin, known as the antielectron (also called the positron). The proton has a hidden twin called the antiproton, and the same holds for all other particles. *But the antimatter counterparts of our regular matter particles are almost nowhere to be seen in everyday life.* Where they are, and why they disappeared shortly after the Big Bang created them, is one of the greatest mysteries in physics, and in this chapter we will learn more about it and about the LHC's search for an answer.

Paul Dirac was the physicist who came up with the outrageous idea that nature must also contain antimatter: particles like the ones we observe, but with opposite electric charge. These are weird denizens of an antiworld, which, once they come in contact with ordinary matter, immediately annihilate it, resulting in the release of energy; ounce per ounce, this energy would make even a very large atom bomb pale by comparison.

Dirac conceived the strange idea of antimatter once he derived his own equation, now called the Dirac equation, which described the behavior of tiny particles in a more complete way than ever before. Dirac knew that the Schrödinger equation of quantum mechanics accurately described the behavior of a particle, such as an electron, as long as it was not moving too fast—that is, as long as its speed did not come anywhere close to the speed of light. On the other hand, Dirac knew that for large bodies moving at speeds close to that of light, Einstein's special theory of relativity provided the right approach and could successfully describe their laws of motion. So what Dirac wanted to do was to put together an equation that would combine *both* elements: quantum mechanics (the set of laws of behavior of the very small) and special relativity (the theory of the very fast).

It was a great idea, but it seemed impossible to implement. Other physicists had reasoned along the same lines and had tried to achieve a description of nature that was both quantum based and relativistic. But they had all failed. These scientists included Schrödinger himself. But Paul Dirac was a very unusual scientist.

As an undergraduate, Dirac studied at the University of Bristol. He was interested in mathematics, and he studied hard. The Nobel physicist Eugene Wigner of Princeton University, whose sister Margit ("Manci") had married Dirac, interviewed him in 1962 about his early life. "In the mathematics course," Dirac remembered, "the class was very small—it was only two people, a girl and myself." Wigner asked: "Did you talk to her about mathematics, about the subject?" Dirac responded: "No, I think we only met at the lectures and separated afterward, so far as I know." Wigner teased him: "You didn't invite her to tea?" Dirac answered: "Oh no. In fact, I had no social life at all."[1]

It was little wonder, for Paul Adrien Maurice Dirac was not a talkative man. Two stories about him demonstrate this tendency. When Dirac once gave a lecture at the University of Toronto, someone in the audience raised his hand and said, "Professor Dirac, I don't understand how you derived the equation on the right." Dirac stopped but gave no answer. After a few moments of thought he responded: "This was a statement, not a question," turned back to the board, and continued his talk. Another time, at a dinner in his college at Cambridge University, Dirac was seated next to the novelist E. M. Forster, whose books he admired and who apparently was one of few people as untalkative as Dirac. After a long period of silence, Dirac finally turned to Forster and said, in reference to the latter's novel *A Passage to India*, "What happened in the cave?" Forster did not answer. When the dessert arrived, he finally turned to Dirac and said: "I don't know."[2]

While still at Bristol and not yet hooked on physics, Dirac took a course from a professor named Broad, who brought Einstein's ideas to his students. "When the war ended, there was tremendous interest in relativity," Dirac recalled. "Previously we just hadn't heard about it at all." This was because in 1919 the great English astronomer and physicist Arthur Eddington went to Principe Island in the Atlantic to observe a total solar eclipse; the photographs his team took of stars grazing the eclipsed Sun showed clearly—when compared with a star map—that the Sun bent the light rays from these stars (because mass curves space, as explained by general relativity) in exactly the way Einstein had predicted. Following Eddington's discovery, Einstein became a global celebrity, and continued to be one throughout his life.

Professor Broad taught philosophy, but he gave one course on relativity, and Dirac attended it because he was eager to learn about this new and exciting theory. Unfortunately, however, according to Dirac, "he talked about it largely from the point of view of a philosophy. I tried to appreciate it, but I did not get very much success in trying to appreciate philosophy." But Dirac did learn from Broad about Einstein's theory of special relativity. "I saw that it was really something new which I had never thought of in my speculations about relations between space and time."[3] Dirac, who until then was unsure of what he wanted to study, decided at once to apply to graduate school in physics at the prestigious Cambridge University.

At Cambridge, Dirac learned about Bohr's theory of the atom. He had not realized how far developed atomic theory was, having spent most of his time studying mathematics. However, Dirac now understood the theory to the point where he could actually start to contribute to it. He began to work in earnest and became obsessed with the idea of wedding quantum theory to special relativity in order to describe the nature of small particles in a far more complete and satisfactory way than had been possible until then.

Dirac spent many hours a day in contemplation, taking long walks in which he would concentrate on the problem, and worked nights with paper and pencil. But he was getting nowhere. His starting point was the Klein-Gordon equation, which was supposed to do the trick but always led to absurd results. Dirac tried to alter the Klein-Gordon equation in order to make it correct but kept failing no matter what he did. He was getting very frustrated and depressed.

Then, one cold evening in 1928, something happened. Dirac was sitting in front of the fireplace in a lounge at Cambridge University's Saint John's College, staring into the fire. And then suddenly he saw it. Dirac realized that the Klein-Gordon equation was flawed because it contained a term describing how the system changed with time that was doubly applied. (Technically, the Klein-Gordon equation had a second-order derivative in time.)

Dirac saw that the equation he was after must apply change with time only *once* (meaning it should contain only the first derivative with respect to time)—as does the original Schrödinger equation. He thus discarded Klein-Gordon and went back to Schrödinger's work. To apply the prescriptions of special relativity to quantum mechanics, Dirac invented four *matrices* (these are arrays of numbers used in mathematics, for which certain rules for multiplication, addition, subtraction, and inversion apply), which would later become known as the Dirac gamma matrices. The gamma matrices enforce the requirements of special relativity—the fact that speeds cannot exceed that of light—and once one of these matrices was cleverly incorporated into the framework of quantum mechanics, everything fell into place as if by magic: The Dirac equation was born.[4]

Dirac's equation was shown to lead to correct predictions about the behavior of small particles, even those moving at speeds close to that of light, and to provide solutions to problems in particle physics that were proved to be far more accurate than anything known until that time. The infusion of special relativity into the quantum world through the Dirac equation also incorporated Einstein's famous relationship between mass and energy, $E = mc^2$.

When Dirac analyzed his own equation—reportedly on the day after he derived it—he found a surprising property. His equation, when applied to an electron, contained a specific term that he understood to represent the electron's spin. The idea of spin, until that time only inferred experimentally, thus emerged directly from Dirac's powerful equation. The spin of the electron, and of all other elementary matter particles (fermions), is defined as ½, in units of Planck's constant.[5] Bosons (a category that includes the force-carrying particles) have spins that are integers: 0, 1, or 2, in terms of Planck's constant. For example, the photon, which is the carrier of the action of the electromagnetic force, has spin 1.

When Dirac solved his equation for the electron, he quickly noticed another interesting property: Some of the solutions had *negative* energy levels, which he called "holes." What is *negative* energy? It seems to make no sense at all. It's as if you woke up in the morning, worked out on the treadmill for a while, and suddenly the number of calories burned, shown on your digital display, started *decreasing:* from 85 to 84 to 83. . . . Wouldn't this be crazy? But that's exactly what would happen if your body started generating *negative* energy; when added to positive energy, the total calories expended would go down.

Many other physicists, when faced with something like this, an idea that made no sense to them, would simply assume that there was something wrong with the derivation or at least that the negative-energy solutions they had obtained were just a mathematical oddity. Not so Paul Dirac. He was convinced that his new equation had to be telling him something real and important about the universe. Using the symmetry inherent in negative and positive energy levels, Dirac concluded that an electron with a negative energy level is *another kind of particle*—one that is the "opposite" of the electron!

At first he thought that the proton might be that electron with the negative energy—but the problem of mass immediately arose: The proton is far too heavy to be an electron with negative energy. Besides, if the proton were such an animal, it would annihilate with an electron in an atom, and matter could not exist. So Dirac had to think some more about this puzzle, and his conclusion stunned the world of science. Dirac predicted the existence of a new kind of particle: the antielectron, now called the *positron*. It is a particle with the same mass as a regular electron, but with identical and *opposite* electric charge. The positron has a *positive* rather than a negative charge of one unit.

In 1932, Carl D. Anderson of Caltech confirmed the existence of the positron while studying cosmic rays that entered a cloud chamber—a device that detects particle tracks by displaying the condensation of mist formed when a particle passes through it. The trajectory of the particle in a magnetic field in the cloud chamber was the opposite of that of the electron, and the mass was inferred to be identical to it: The particle had exactly the properties of the positron predicted by Dirac. In 1933, Paul Dirac shared a Nobel Prize in physics with Erwin Schrödinger. Anderson himself received a Nobel Prize in 1936 for discovering the positron and for further work showing that electron-positron pairs were produced by gamma rays.

Dirac's equation brought into physics the whole idea of *creation* and *annihilation* of particles. Particle-antiparticle pairs can be created by nature out of sheer energy. No one had conceived of such weird phenomena before; particles, like everything else in the world of physical objects, had been thought to simply *be*. Now nature's way of creating particles was beginning to be understood.

Dirac had gone way out on a limb as a physicist, taking great risks with his reputation right after having achieved a monumental theoretical breakthrough in wedding special relativity and quantum mechanics. Instead of stopping there, he made an outrageous prediction: the existence of the bizarre antiparticles. But his gamble worked: Nature does act in strange ways and produces not only matter but also antimatter.

Dirac later further deduced that all the particles of nature should have twins that were the opposite of themselves, just as in the electron-positron case. The neutron would not be discovered by James Chadwick until the following year, so the only remaining known particle at that time was the proton; Dirac concluded that it had to have a twin, the antiproton, a particle with the same mass as the proton but with a negative charge of one unit. In 1955, the Italian physicist Emilio Segrè and the American physicist Owen Chamberlain discovered the antiproton in work done at the University of California at Berkeley. They shared a Nobel Prize in 1959.

Dirac's antiparticles brought a new element to the world of physics. Every particle has its own antiparticle: a twin that has the same mass, but opposite electric charge. Charge is conserved in particle interactions, so if a neutral particle decays into a positively charged particle, then a negatively charged one (of equal amount of charge) must also be produced so that the total charge before and after the reaction remains zero—the charge of the original, neutral particle. Remember the other important fact, due to Einstein: Energy and mass are the same. This means that energy can spontaneously turn into mass, as we have seen before, but charge must be conserved in all such processes. These requirements lead to the production of particle-antiparticle pairs.

The vacuum of space is now understood not to be empty: In fact, it is teeming with activity. According to the laws of relativity (the interchange of mass and energy) and quantum mechanics, energy can suddenly and with no apparent reason produce pairs of particles: electrons with their opposite twins, positrons. Such reactions are indeed observed in the vacuum of space: Pairs of particles can be seen through the tracks they leave in cloud chambers and similar detectors. Electron-positron pairs appear out of nowhere in space, and they disappear just as quickly, each annihilating its antipartner.

The Big Bang launched our universe 13.7 billion years ago as a powerful burst of energy. And, as we have seen, energy can turn spontaneously into particle-antiparticle pairs. This is why scientists believe that the Big Bang must have created equal amounts of matter and antimatter in the primordial soup of elementary particles that followed. When these particles and antiparticles met, they mutually annihilated, turning back into energy, which then again produced pairs of particles and antiparticles, and so on. At some point, however, this process must have stopped, because today we live in a stable universe that doesn't self-annihilate.

Somehow matter won over antimatter, allowing our matter-dominated universe to emerge. For while equal amounts of matter and antimatter may have been created in the Big Bang, as many scientists believe had to have happened, matter survives in our universe, and antimatter can only be observed in cosmic rays, in certain radioactive decay processes, and when it is artificially made in particle collisions in accelerators or in the medical diagnostic machine called a PET (positron emission tomography) scanner.

Scientists want to know what caused the victory of matter over antimatter, since on the face of it, both kinds of matter seem to be perfectly, symmetrically the opposite of each other. Since they mirror each other, they might be expected to behave in essentially the same way. But in what way? Well, let's reverse all the electric charges of the *matter* particles in the universe. This should give us the antiparticles. But there is one more thing to do. Remember that a positron (which carries a positive electric charge) curves in a magnetic field in the *opposite* direction to the curve of the track of an electron (which is negatively charged). So if we reverse the electric charge *and* reflect the original particle's track in a *mirror*, we should see exactly the same physical behavior for antiparticles as we do for particles. Combining both charge reversal, which we'll denote by C, and reflection in a mirror, called *parity inversion*, which we'll denote by P, should give us the same physics. If you perform C and P on an antiparticle, it should look exactly like a particle; and vice versa. This idea is called *CP conservation.*

Left-right symmetry in nature (reflection in a mirror) goes with a presumed conservation law called *parity (P)*. Until the middle of the twentieth century, physicists believed that parity itself was conserved in nature, meaning that if you reflect the world in a mirror, you get exactly the same physics. Then the surprising truth came out.

One day in early May 1956, the Chinese American theoretical physicists C. N. Yang, then of Brookhaven National Laboratory, and T. D. Lee of Columbia University were dining together in a Chinese restaurant in New York. They were discussing parity and together developed the suspicion that it may be conserved everywhere in physics *except* in a certain kind of radioactive process called beta decay, which is governed by the weak nuclear force. This means that the mirror world acts the same as our world except when acted on by the weak force. The pair of physicists then suggested several kinds of experiments that could be used to test their hypothesis. Yang and Lee published a paper with their theory in the journal *Physical Review,* but it was not received with enthusiasm by other physicists. People did not want to believe that nature may somehow break one of its presumed symmetries.

Frustrated by the cool reception to their ideas, Yang and Lee sent their paper to Wolfgang Pauli, hoping to get his support, but reportedly Pauli threw the paper away.[6] He did not believe that mirror symmetry was violated anywhere in the universe and did not see why the weak force would act any differently from the other three forces of nature. When the MIT physicist Victor Weisskopf (later to become director general of CERN from 1961 to 1965) asked him about the new hypothesis, Pauli wrote him: "I do not believe that the Lord is a weak left-hander"—"weak" here refers to the weak force, and "left-handedness" refers to the violation of parity. Pauli then added that he would bet a large sum of money against such a finding.[7]

Experimental physicists were equally reluctant to search for the parity violations suggested in the Yang and Lee paper because such tests would be difficult to carry out, requiring the use of highly radioactive elements and very precise measurements with such materials. But somehow Yang and Lee managed to convince a female fellow Chinese American physicist, Chien-Shiung Wu, affectionately known by physicists as Madame Wu, a Shanghai-born experimental physicist working at Columbia University, to carry out an experiment on their behalf.

Wu and her colleagues proceeded to design and conduct a test with atoms of the radioactive element cobalt 60 undergoing beta decay, whose nuclei had been aligned in a magnetic field so that their spin directions were uniform. In beta decay, an electron is released from the nucleus. Wu's experiment was aimed at finding out whether the electrons were released randomly, or whether one direction was preferred over another. If there was a preferred direction, then mirror reflection symmetry, or parity, would be violated. In the experiment, the cobalt atoms were cooled to a very low temperature to keep them from vibrating, and a magnetic field was applied to control the spins. Now, if parity was conserved, then the process of emitting an electron would look the same if it was viewed in a mirror.

When the directions of the electrons emitted in the beta decay were studied, the researchers were surprised to find that there was no symmetry in directions: One particular direction was preferred over its mirror image. This finding implied that parity was not a conserved quantity—that the mirror image of a weak-force decay process was not the same as the original process, exactly as Yang and Lee had predicted would happen. Wu's successful experiment led to the 1957 Nobel Prize in Physics for Yang and Lee, who got their awards in the record time of less than one year. However, many physicists expressed disappointment that Wu did not share the prize.

When Pauli received the news about the experiment, he felt humiliated. It seemed that his keen physical intuition had failed him. He wrote Weisskopf that he now accepted that the Lord was a "weak left-hander" after all, and added that he had been "very upset and behaved irrationally for quite a while" after hearing the news about Wu's results.[8] He then sent Bohr an "Obituary to our dear female friend of many years, PARITY, who had gently passed away on January 19, 1957, following a brief suffering caused by experimental treatment. Signed: electron, muon, and neutrino."[9]

Similar experiments to Wu's were carried out by Richard L. Garwin, Leon Lederman, and Marcel Weinrich, and also by Jerome Friedman and Valentine Telegdi with muons—the heavier cousins of electrons. Both studies, published in 1957, found that the electrons that came out of these reactions also had a preferred direction, confirming the parity violation discovered by Madame Wu and her team. On February 15, 2010, scientists from Brookhaven National Laboratory reported that the quark-gluon plasma they had created by smashing together gold atoms at 99.995 percent the speed of light showed tiny pockets in which parity was violated. It seems that at the immensely high temperatures of the early universe, around 4 trillion degrees Celsius, which the Brookhaven accelerator had produced, the behavior of quarks and gluons also does not respect mirror symmetry.[10]

But parity violations alone cannot explain the disappearance of antimatter from the universe; other experiments over the years have shown that the world is not the same if electric charges are reversed. But the salient question is whether *both* parity and charge reversal are violated at the same time, because this would imply that matter and antimatter *are* indeed different: they don't behave in the same way. Therefore, what we need in order to discover a fundamental difference in behavior between matter and antimatter is a *CP violation.*

In 1964, James Cronin and Val Fitch, working at Brookhaven National Laboratory, showed that the decays of certain mesons, called neutral kaons, violated CP symmetry, demonstrating that antimatter does not act exactly like matter. The degree of this discrepancy, however, was very small—the kaons broke the CP symmetry only 0.2 percent of the time. Such a small difference is not enough to have caused the cosmic disappearance of the antimatter.

The LHCb collaboration at CERN uses a special-purpose detector to look for processes that violate CP symmetry to a higher degree. If significant differences in behavior do occur, more than the small violation already known, they could explain why matter won over antimatter in the early universe. But the LHCb project is only the latest in a long line of experiments at CERN aimed at uncovering the nature of antimatter. The laboratory began to study the mystery of antimatter fifteen years ago, when CERN scientists commenced the first important experiments to create atoms of antihydrogen.

CERN's Low Energy Antiproton Ring (LEAR) was used to produce antiprotons, which were then aimed at a target of xenon gas. In the resulting collisions with xenon nuclei, positrons were created, and some of them joined antiprotons to make antihydrogen atoms—the first atoms of antimatter ever created by humankind. Once antihydrogen atoms are made, they are no longer charged (whereas the individual positrons or antiprotons are). They therefore can no longer be magnetically confined to the center of a vacuum-filled trap, and they drift to the edges. There they meet regular matter and annihilate (because the positron annihilates when it meets an electron from matter, and the antiproton annihilates when it meets a proton in any nucleus of matter). The amounts are so small, in comparative terms, that the energy produced is tiny and therefore does not cause any damage. But while these antiatoms float freely, before they self-destruct, scientists can study their behavior.

Since hydrogen is the most common element in the cosmos, making up about 75 percent of all matter in the observable universe around us, comparing its behavior with that of antihydrogen should prove useful for determining why matter is prevalent in the universe. Scientists study the orbits of the positrons in the antihydrogen atoms, determine their energy levels, and compare them with those for the electrons in normal hydrogen atoms. (The energy levels of hydrogen have been studied extensively for many decades, so we know a lot about them.) Scientists also want to know if gravity somehow acts on antimatter differently from the way it acts on matter. If so, this may help solve the mystery of antimatter and explain why we exist.

Scientists at CERN and elsewhere are studying antimatter because they are looking for an answer to the mystery of the disappearance of antimatter after the Big Bang, which supposedly led to our matter-dominated universe. But what if this thinking is wrong? What if, instead of the perceived tremendous imbalance between matter and antimatter, there is actually a whole mirror universe out there in which antimatter dominates over matter? Or what if we live in a single universe that is a patchwork of nonoverlapping smaller realms alternately consisting of matter and antimatter, separated enough from each other so that they don't mutually annihilate? If we live in either kind of such a universe, then all the conclusions we've been entertaining are wrong, and matter never did win over antimatter—they just went their separate ways.

The idea that antiworlds exist in which everything is made of antimatter became popular in the 1960s, inspired, in part, by the book *Worlds-Antiworlds* by the Nobel physicist Hannes Alfvén. He believed that some of the stars we see in the sky are made of antimatter, and that these stars and planets are separated from our world by lots of empty space, leading to very little mixing, which prevents mutual annihilation.

This notion inspired George Smoot to pursue a scientific project in search of antiparticles arriving in our stellar neighborhood. Smoot and his colleagues at the University of California at Berkeley lofted large balloons to the edge of our atmosphere, where the air is very sparse and particles from space could readily impact the research team's airborne detector. In particular, Smoot's group was looking for actual *nuclei* of antimatter: composite particles made of several antiparticles combined together. The assumption was that such particles would have to have emanated from an antiworld, rather than from collisions in our world, since such collisions, or radioactive processes, can produce only small and simple (rather than composite) antiparticles.

To guide their exploration, Smoot's team designed a sophisticated detection system that used liquid helium to cool a superconducting magnet, creating a powerful magnetic field to curve strongly the path of an incoming particle entering the detector. If a particle was found to leave a track that curved in the *opposite* direction from that of a normal particle for a given mass (weight was gauged against those of known atomic nuclei), it would constitute strong evidence that it was made of antimatter.[11]

The team carried out experiments with high-flying balloons for several years. At one point, the researchers made an exciting discovery: A particle left a strange track that was the opposite of what would have been expected of a matter particle. By its weight, it looked like an oxygen nucleus, but the track it left in the detector curved in the opposite direction. The team named this unusual particle "cosmic ray event 26262," labeling it by the number of the segment of film on which it was recorded. (This was one of fifty thousand events recorded on hundreds of feet of photographic film.)

Was this discovery a real antimatter nucleus—the central component of an actual antimatter atom that came here from outer space? Or was it just an error in the data recording? This question consumed Smoot and his team members. Eager to resolve the conundrum, they performed intensive data analyses using advanced statistics, which revealed that the probability that "No. 26262" was an actual antioxygen ion was about 75 percent, which would appear excellent. But the team, advised by the legendary Berkeley physicist Luis Alvarez—who had proposed the now widely accepted theory that the dinosaurs became extinct 60 million years ago because of the catastrophic impact on Earth of a comet or asteroid—told them that 75 percent certainty was not a high enough level of confidence. Reluctantly, Smoot and his colleagues wrote off the event as a fluke and ended their project.[12]

In 2009, the British physicist Frank Close of Oxford University published a small book titled *Antimatter,* aimed at dispelling the myth that grew out of the Dan Brown book and film *Angels and Demons*— that antimatter can be used to create weapons of mass destruction. But before demonstrating that production of enough antimatter to make such a bomb would be prohibitively difficult, would consume untold resources, and would take many thousands of years, Close tells a story about a cataclysmic event that took place a century ago.[13]

At 7:17 a.m. on June 30, 1908, a tremendous explosion was heard, and the daylight sky was lit by a fireball that made the sun pale in the desolate region of Tunguska, Siberia, more than 500 miles from the nearest city. Forests in the entire 800-square-mile area of the Stony Tunguska River burned, and thousands of trees were blown away by the force of the blast. The few farmhouses in this sparsely populated region were very badly damaged, and silverware in some destroyed homes was melted from the heat.[14] Entire herds of reindeer were vaporized, and the forests did not recover from the damage of the blast for more than thirty years.

For several months after the explosion, the skies over Europe were luminescent at night—people recalled they could read by the hazy light reflected from the night sky. It is now estimated that the force of the explosion was about 15 *megatons*—a thousand times more than the power of the bomb that destroyed Hiroshima. The blast was so powerful that, had it taken place over Chicago, it would have been heard over much of North America, and the explosion's flash would have been visible from the southern United States to much of Canada.

But this immense explosion—which took place almost four decades before nuclear bombs were invented—had one additional bizarre characteristic: It left no visible crater. This meant that no appreciable amount of hard matter hit the ground, as happens when an asteroid or comet strikes the Earth, or as we see on the crater-riddled Moon.[15]

In 1965, the American Nobel Laureates W. F. Libby, who had invented the radiocarbon dating method, and Clyde Cowan, the codiscoverer of the neutrino (who received a Nobel Prize posthumously in 1995, two decades after he died), and the scientist C. R. Atluri published a joint paper in *Nature* in which they proposed their theory that Tunguska was caused by the annihilation of a piece of antimatter hurled from space. They used evidence from radiocarbon analysis of trees, whose rings had been counted to determine the exact date, showing that during the year following the disaster, 1909, there had been an unusual abundance of radioactive carbon in our atmosphere. The three scientists argued that this result was consistent with the radiation that would have been produced from the mutual annihilation of matter and antimatter.[16]

Eventually the theory was questioned by other scientists, and the leading view today is that an asteroid or comet was the real culprit; some have even tentatively identified a crater some distance away from where it might have been expected.[17] But nobody knows for sure what happened at Tunguska.

The Alpha Magnetic Spectrometer (AMS) is a large superconducting magnet detector, similar to but much smaller than those of the LHC, designed to be attached to the International Space Station. The AMS project is aimed at studying cosmic rays, searching for dark matter candidates, and looking for traces of antimatter. The project of launching this expensive magnetic detector into space, with its heavy supply of liquid helium, demonstrates that at least some scientists still seriously consider the possibility that stars and maybe even planets made of antimatter may exist somewhere in the universe. They continue to look for evidence for the existence of antimatter, or, if it is not found, to eventually rule out the possibility. An older satellite, called PAMELA—Payload for Antimatter Matter Exploration and Light-nuclei Astrophysics—was launched by the European space agency in 2006, in part to look for antimatter particles arriving from space.

Many other scientists, however, are working under the assumption that all the stars and galaxies we see in the heavens are made of matter and not antimatter. If, indeed, everything is made of matter, and if the Big Bang really produced equal amounts of matter and antimatter, then where *is* the antimatter? Where did it go after the Big Bang?

The special-purpose detector LHCb—where "b" stands for "beauty," in reference to the heavy *beauty* or *bottom* quark—searches for an answer to the question: Is antimatter significantly different from matter to the point that matter could have won over antimatter? The collaboration looks at decays of B-mesons in search of substantial CP violations. Unlike the other three main detectors of the LHC, the

LHCb detector is located close to the path of the incoming colliding protons because the b-quarks and anti b-quarks travel close to the original beam, rather than in a wider range of directions.

When the universe was only a hundredth of a billionth of a second old, very heavy kinds of quarks and antiquarks were circulating in it, annihilating each other when they met (as quark-antiquark pairs). These heavy quarks are called b-quarks, or bottom quarks, or beauty quarks. These quarks also live inside heavy mesons, called B-mesons. The decay products of the high-energy proton collisions inside the LHC include a certain number of b-quarks and anti b-quarks found inside the B-mesons the machine produces. The LHC thus mimics the creation of these particles and antiparticles after the Big Bang. The LHCb detector analyzes the behavior of these particles and their interactions, and the scientists in this collaboration search through these data hoping to detect a larger degree of asymmetry between matter and antimatter than is presently known, so we can learn why matter won—if indeed it had—and why we are here and able to ask such questions.

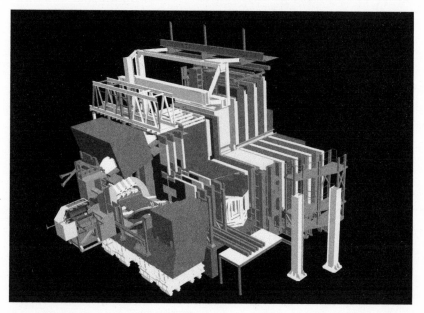

A schematic diagram of the LHCb detector

Chapter 6

Richard Feynman
and a Prelude to the
Standard Model

When Paul Dirac wrote his equation incorporating special relativity with quantum mechanics, he launched *quantum field theory*. As a result, in modern physics, particles and their interactions are now viewed in the context of *fields*. The first fields to be studied were the electric and the magnetic fields, unified through the work of the Scottish physicist James Clerk Maxwell in the nineteenth century and now called the electromagnetic field.

The idea of a field becomes tangible if you consider just the magnetic field, which we all know from games we played as children. If you place a bar magnet under a piece of paper and sprinkle iron shavings over it, you will immediately see that the particles of iron align themselves in a pattern from one edge (pole) of the magnet leading to the other. What you are looking at is a depiction of the magnetic field created by the bar magnet.

When Einstein was a child, his father gave him a present: a magnetic compass. The boy spent many hours marveling at the unseen force that led the compass needle to align itself with Earth's magnetic field, pointing north to the magnetic pole. As a young scientist some years later, Einstein used the idea of fields in his work. His general theory of

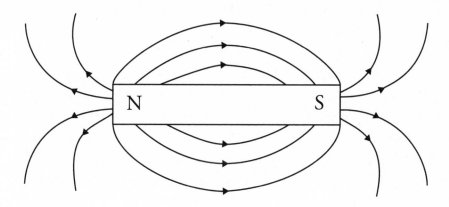

relativity—about the *gravitational field* created by massive objects—is one of the most important field theories we have. Following Einstein's lead, the most useful theories of physics today are field theories.

Quantum field theory is the modern theory of particle physics, based on the idea that fields are the essential elements in nature. Here the fields involved obey the laws of quantum mechanics—the super-position principle, Pauli's exclusion principle, Heisenberg's uncertainty principle, particle-wave duality, quantization—and also those of Einstein's special theory of relativity, which ensures that nothing can go faster than light, and that mass and energy are equivalent.[1]

Fields can get *excited*. Think of a field as a mattress, and of someone jumping up and down on it. This excitation of the field creates a *wave*—a vibration of the springs that propagates through the mattress. Now according to quantum mechanics, waves are particles and particles are waves, so the excitation of a field thus creates a *particle*. But in quantum field theory, we also have from special relativity the property that energy and mass are equivalent, and this equivalence allows the energy produced by the field excitation to be potentially transformed into mass, hence giving us *massive* particles.[2] Unlike quantum mechanics alone, quantum field theory can actually *create* and *destroy* massive particles. In the LHC, the particle collisions create energy, which excites fields, creating massive particles. The greater the energy the machine generates, the more massive the particles that can appear.

The elementary particles we know are all members of the *Standard Model.* As mentioned earlier, the Standard Model is a very successful theory in physics, a quantum field theory that makes excellent predictions about particle behavior. This model was constructed over a period of decades. The first successful quantum field theory was created independently by Richard P. Feynman, Julian Schwinger, and Sin-Itiro Tomonaga. It is the theory of quantum electrodynamics, which describes the interactions of electrons with photons. This was a very important theory because it explained how a ubiquitous matter particle, the electron, interacts with the electromagnetic field through the action of a force-carrying particle—in this case, the photon, our usual ray of light. This theory would pave the way to the Standard Model of particle physics, which includes all the known matter particles (and antiparticles) and the known force-carrying particles (which are bosons).

Richard Feynman was an American physicist who in his lifetime attained celebrity status. Descriptions of his adventures—from bongo drum playing to safe cracking, to hypnotism, to topless-bar hopping—are familiar to many through his popular books and lectures. But interviews contain even more fascinating details.

When Feynman was a baby, his father sat him on a high chair and placed before him bathroom tiles of different colors, asking him to arrange them in set patterns: a blue followed by a white, followed by a blue. From that early age the boy began "to think about patterns and recognize those things as being interesting. After a short time with this game, I could do extremely elaborate patterns." When he got older, he noticed a book about algebra on his father's bookshelf. He asked his father what algebra was.

He said that it had to do with doing problems. This I remember exactly: "It's a way of doing problems that you can't do in arithmetic." I said: "Like what?" He said: "Like a house and a garage rent for $15.00. How much does the garage rent for?" I said: "But you can't do that at all!" He left me with that answer. He didn't tell me what algebra was.[3]

Feynman's early interest in mathematics continued, and as an undergraduate he went to MIT to study it. But soon he found pure mathematics to be too abstract to hold his interest, and he went to talk to the head of the department. Feynman asked him what was the use of higher mathematics "other than to teach more higher mathematics." The department chair found this comment irritating and replied that someone who asked such questions was probably not suited for mathematics, so after considering a major in engineering, Feynman turned to physics. And here he found his true calling.

Once when Feynman came home for vacation, his father asked him about his studies.

"Well," he'd say, "young man, I helped you get started in science and sent you to MIT to learn something, so you should come back to your Old Man and teach him something," he said, "There's something I never understood that I want you to explain to me."

I says, "What?"

He says, "They talk about an atom in an excited state emits a photon, which is like a particle."

I says, "Yeah."

He says, "Now the particle is not in the atom ahead of time—huh? And is not in the atom afterwards—one less photon—huh? It just comes out, this particle? Explain that to me, please."

I says, "Father, I cannot!"

He says, "I've been frustrated. All these years I worked—!"

I tried my best. I said, "It's like sound coming out of a box. The sound is not there ahead of time, but it comes out—"

He said, "Well, it's the energy of the vibration?"

"Yeah, well, it's the energy that comes out in the form of the photon."

He says, "Yeah, but the photon is a particle, is it not?"

I say, "In certain ways, yes, in certain ways—"

He says, "Come on, now!"[4]

But it would be Richard Feynman who would give us a complete understanding of this exact phenomenon, in the form of the theory of quantum electrodynamics.

In 1939, after graduating from MIT, Feynman continued on to a doctoral program in physics at Princeton, working with the renowned physicist John Archibald Wheeler. Feynman wrote a brilliant dissertation under Wheeler, extending the early results of Paul Dirac, and creating the sum-over-histories approach in quantum mechanics. This method considers all possible paths that a particle can take when going from one point to another. Each path has its own probability, and there are ways of aggregating the information to learn about a process.

A few years ago, when I interviewed him about the work Richard Feynman had done under his direction, John Wheeler recalled that he was so excited about his student's work that he rushed to Einstein's house, 112 Mercer Street in Princeton, to show it to the great man. "Isn't this wonderful?" Wheeler asked Einstein, as the latter was perusing Feynman's thesis. Einstein had, of course, famously opposed quantum mechanics and its use of probabilities, of which Feynman was now making even greater use in his thesis. Einstein looked up from the manuscript, contemplated the question for a moment, and responded: "I still don't believe that God plays dice. But maybe I've earned the right to make my mistakes."[5]

Not long after he finished his thesis, Feynman was preparing to give a talk about his work at the physics department colloquium. Einstein came in while Feynman was filling the blackboard with formulas.

Einstein looked at the board, and then asked him: "Where is the tea?" Feynman directed him to the tearoom, where professors and guests often went before the presentations. Then Wolfgang Pauli, who was visiting Princeton at the time, came in, and he too went to have tea. Afterward, Feynman's lecture started.

Feynman described what happened after he had finished his talk.

> Professor Pauli got up immediately after the lecture. He was sitting next to Einstein. And he says, "I do not think this theory can be right because of this, that, and the other thing. . . ." It's too bad that I cannot remember what . . . the gentleman may well have hit the nail on the bazeeto, but I don't know, unfortunately, what he said. I guess I was too nervous to listen, and didn't understand the objections. "Don't you agree, Professor Einstein?" Pauli said at the end of his criticism. Einstein said "No," in a soft German voice that sounded very pleasant to me.[6]

Then Wheeler stood up and explained all the points that remained unclear, answering Pauli's objections. It was evident that what Feynman had achieved was a great theoretical advance in physics. He had developed a key tool for calculating previously intractable quantities in quantum mechanics, which would find continuing applications over the decades to follow. Computations needed in particle interactions often use Feynman's tools. But he would later do much more.

During the war, Feynman worked on the Manhattan Project, and when he returned to academia at war's end, he was offered a professorship at Cornell. When he first came to Cornell, he slept in a lounge. "Then suddenly I realized: I am a professor!"[7] In fact, he became such a popular professor that he stopped doing physics research altogether and concentrated on his teaching. Theoretical work had become uninteresting to him.

Then one day, a chance event brought Richard Feynman back to the frontiers of physics.

I was in the cafeteria eating as usual. I used to eat in the students' cafeteria because I liked to look at the girls. And some kid throws a plate up into the air. You know how kids are. Now the plates at Cornell had a blue seal at one side of the rim, and he threw his plate up in the air, and it was sort of flat and wobbled, almost horizontal but with a slight wobble. At the same time, the blue mark on it, the insignia on the plate, went around the plate. The wobble and the motion seemed to be related. So I wondered, what is the relation? How many wobbles per rotation is it?[8]

Feynman studied the problem mathematically, and he discovered that when the wobble wasn't very severe, the insignia went around twice as the wobble motion completed one turn. He thought this was "cute, and it's a nice relationship, two to one."[9] But he wanted to learn more about the theoretical process behind this result.

So after considerable effort that afternoon, working with the equations, drawing diagrams and showing the forces at play, Feynman saw a way to explain the motion. Having done it, he ran into department chair Hans Bethe's office and said: "Hey, I saw something funny about a disk," and explained it to him. Bethe responded: "But what's the importance of that?" Feynman said: "Hans, it doesn't have any importance. I don't care whether a thing has importance. Isn't it fun?" Bethe said: "It's fun."[10]

But Feynman did find an important use for the property he had discovered. The spinning, wobbling dinner plate reminded him of old unanswered questions in physics about the spinning electron, and about how to represent the spin in his own invention, "path integrals" in quantum mechanics. This brought Feynman back into the fold of serious research work in physics, rather than just play. "It just opened the gate," he recalled.[11]

The spin of the electron had to be incorporated into Feynman's earlier work at Princeton under Wheeler. But when he extended these results to higher dimensions, infinities appeared—the plague of particle physics—and the need arose for *renormalization*: the theoretical process of removing unphysical infinities from mathematical calculations.

In 1946, Feynman was invited to speak at the bicentennial conference at his alma mater, Princeton. It was an unusual meeting in that it included both scientists and high school teachers. There, Paul Dirac was a speaker, and Feynman presented him. After Dirac spoke, Feynman tried to explain Dirac's highly technical presentation to the high school teachers, but his explanation fell flat.

As it turned out, Niels Bohr was in the audience. Feynman described what happened next:

Then Bohr got up and said, "Feynman makes very many jokes," and so on, "but aside from the jokes we have some important problems here to discuss." . . . And then he made some commentary which I thought was absurd, which was that the infinities that we were getting in these various theories were all going to balance out, that there would be more particles—there's protons, then there's mesons, there's this, there's plus signs and minus signs and plus signs and minus signs, plus infinities, positive energies and negative energies . . . and they're all going to add up, so there was no problem.

That sounded crazy to me, just instinctively. It was, of course. So I didn't like that theory. And they're all sitting around worrying, and they're talking, and I look out the window, and through all this Mr. Dirac, paying no attention to anybody, had walked out and was sitting on the grass, lying on the grass with his elbow against his head, looking up at the sky.[12]

Unhappy with Bohr's explanation about the infinities, Feynman continued to work on renormalizing his theory of electrodynamics— attempting to remove the nonsensical infinite solutions of the equations. Hans Bethe and Julian Schwinger were working on the same problem. Feynman developed his own set of rules for computation— which compared favorably with what the other physicists were getting.

Then there was an important meeting of physicists in the Pocono Mountains. Schwinger was going to explain how he did things, and Feynman was to show his own way of performing the same calculations. Schwinger was a highly respected physicist, which made Feynman so nervous he could not sleep the night before his presentation. But once Schwinger and Feynman met and compared their results, they got along very well: each encouraged the other.

At the meeting, while Feynman was trying to explain his work, people kept asking him, "Where does this formula come from?" and "How do you know it gives the right answers?" Only Schwinger understood what Feynman was doing, because he had done the same calculations and derivations. Feynman had invented a new mathematical tool to perform these tasks, which he called "ordered operators." But he had to show how it worked. Then he had to explain how his other method, the path integrals, worked. His methods gave all the right answers in known cases, so he knew that they could be trusted, and he used them to work out new cases.

Paul Dirac was also in the audience, and Feynman recalled that Dirac raised his hand in the middle of the presentation.

> Dirac asked: "Is it unitary?"
> I said, "I'll explain it to you, you can see how it works, and then you can tell me if it's unitary."
> I didn't even know what that meant (is it "unitary"). So I went a little further. You know, we'd get into these arguments. Then Dirac would come up, you know, "Is it unitary?"
> I said: "Is *what* unitary?"

He said, "The matrix which carries me from the present position to the future position."

I said, "I haven't got any matrix which carries me to the future position, I go forwards and backwards and forwards in time. So I don't know."[13]

Dirac was not the only one who could not understand Feynman's innovation. Bohr was at this meeting as well, and he too had difficulty. Ironically, both of these giants of modern physics had themselves developed theories that were so advanced for their time that older physicists could not understand them. Now the two of them were confronted by the groundbreaking work of a young new star.

Feynman had been using what are now called "Feynman diagrams," which show one dimension of space and one of time (since paper is only two-dimensional) and portray the movements of particles and the interactions among particles in a very clear, visual way. Feynman had invented these graphs to explain quantum electrodynamics, but they are exceptionally useful everywhere in particle physics. All the interactions of particles in the LHC collisions, for example, are well represented by Feynman diagrams.

Quantum electrodynamics is the theory of the interactions of

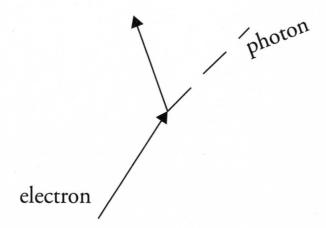

electrons with photons. The first Feynman diagram shown here illustrates a basic interaction in quantum electrodynamics. Here, one electron is moving through space. Note that the horizontal axis represents space, while the vertical one represents time. Then, at a given point in time (the "corner" in the space-time path of the electron), the electron changes direction and emits a photon (the broken line).

The second Feynman diagram shows two electrons moving toward each other. At a given point, they exchange a photon: One electron sends a photon to the other electron, which absorbs it. Each of the two electrons then continues in a different direction.

These two diagrams show the interactions of quantum electrodynamics very clearly. To a physicist, they are also a tool for representing complicated calculations of energies and other parameters of a physical problem, and they simplify the physicist's thinking by making processes visual.

In 1965, Richard Feynman won the Nobel Prize, shared with Julian Schwinger and the Japanese physicist Sin-Itiro Tomonaga, who had also developed theories of quantum electrodynamics, the theory of how light interacts with matter. Physics now had a complete theory of how electrons interact with each other and with photons.

At the Nobel Prize ceremony in Stockholm, Richard Feynman gave

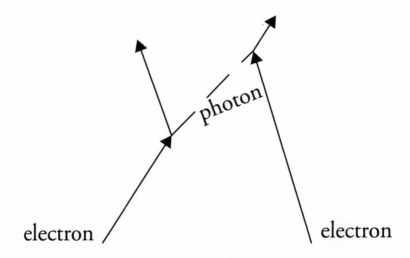

a talk about his work, but he was unhappy with his own lecture, perhaps feeling that the audience didn't understand his theory, which had eluded many able physicists—and this was not an audience of scientists. After the presentations by Feynman, Schwinger, and Tomonaga, there was a formal ball honoring the new Nobel Laureates. So here, Feynman decided to let go and have some fun. He recalled his escapade:

> At the dancing afterwards—you see, I had to get some release from the formalities—I had winked a couple of times at one pretty student. So, when there was a lull, and I saw her again, I went over to her and asked her if she would like to dance. She said, "Yah," and we danced. She danced very well, and we danced kind of wild, and we had a great time. I danced with her quite a lot, to the exclusion of all the formal people, the Princess, everything else. And I must say, it's very amusing, in Sweden it's all completely under control. When I danced with my wife, and when I danced with the daughter of a Nobel Prize winner, they were taking pictures, all the time—click, flash, flash. When I danced with this girl, which I did twice as much as I did with anybody else—no pictures. Nothing. Not in the paper. Apparently they protect the Nobel Prize winners from their own dumb idiosyncrasies.[14]

From there, Feynman went to Geneva, to visit his friends at CERN, where he gave another talk about his work. This one went much better.

> I gave the speech best in front of the group in Geneva, because they're friends. They want to hear what I've got to say. They laugh at the right moment. I mean, you could see in the faces their interest in what I'm about to say, and then the smile when it comes out. There I think I did very much better on that talk than I did in the particular talk in Sweden. I felt good about it too. I was doing all right. I was among friends.[15]

It seems that Feynman's friends at CERN loved his work so much that some of the theoreticians in the lab set out to extend the idea of the very useful tool he had invented, the Feynman diagram. This happened under somewhat unusual circumstances.

One summer evening in 1977, three young CERN physicists went out to a pub after work. They drank beer and told jokes, one of which happened to be about penguins. Everyone laughed, and then one of them challenged the others to a game of darts. Melissa Franklin, now a Harvard professor, started the game with John Ellis, a British particle physicist who is today a leading CERN theoretician. Somehow—still laughing about the penguin joke—the two of them agreed that if Ellis lost the game, his punishment would be to put the word "penguin" in his next published paper. Franklin left the game unfinished and was replaced by Serge Rudaz, who proceeded to beat John Ellis, so strictly speaking Franklin didn't win. Still, Ellis felt he had to live up to his obligation to Franklin to use the word "penguin" in his next physics article.

But how could he do that? He was then working with Mary K. Gaillard, Dimitri Nanopoulos, and Rudaz on the properties of the bottom quark. This fast-decaying elementary particle, the heavier of two "twins" of the down quark that lives inside all protons and neutrons, had just been discovered through a heavy B-meson found at Fermilab. Ellis, Gaillard, and Michael Chanowitz had predicted its mass that spring, before the particle was discovered. So Ellis's next paper would be on the bottom quark. How could a penguin find its way into a paper on quarks?

As Ellis described it later, "For some time, it was not clear to me how to get the word into this b quark paper that we were writing at the time. Then one evening, after working at CERN, I stopped on my way back to my apartment to visit some friends living in Meyrin, where I smoked some illegal substance. Later, when I got back to my apartment and continued working on our paper, I had a sudden flash that the famous diagrams [the Feynman diagrams, here applied to B-meson decay] look like penguins. So we put the name into our paper, and the rest, as they say, is history."[16]

John Ellis left a clue in the paper as to what had really brought him to use the word "penguin"—and thus introduce a new term into the physics literature—by inserting a whimsical acknowledgment in the paper, thanking Melissa Franklin for "useful discussions."[17] The basic Feynman diagram, like the ones shown earlier, looks like a tree (and physicists talk about interactions "at tree level"). But in advanced cases, there is a loop in a diagram. A Feynman diagram with a closed loop and arcs leading into it and emanating out of it might just look like a penguin with a fat belly, wings, and legs. To actually see it, you may have to be in the same state that John Ellis says he was in when he invented it, but see for yourself:

John Ellis drawing one of his penguins on the board at CERN (all the letters represent elementary particles)

Penguin diagrams are especially useful for modeling the decays of B-mesons and therefore hold promise of aiding theoretically in the analysis of the processes studied by the LHCb collaboration at CERN, which, as we recall, looks for solutions to the mystery of antimatter. But, as Ellis pointed out to me, the diagrams are also useful in analyzing phenomena beyond the Standard Model—in particular the search for particles predicted to exist by supersymmetric models.

The theory of Feynman, Schwinger, and Tomonaga explained well the relationship between electrons and photons. The photon in this theory is the boson that mediates the electromagnetic interactions among the electrons. But here, as well as in other areas of particle physics, the values of the constants of nature have not been explained. These constants include the actual masses of particles, which vary widely without our understanding why; and they include other kinds of physical parameters.

For example, the strength of interaction in particle physics is called the *coupling constant,* and it plays an important role in modeling physical phenomena. The coupling constant of quantum electrodynamics is called the *fine structure constant.* Throughout the last century and into the present, this constant has attracted much attention and efforts to "decipher its meaning," because its value happens to be very close to 1/137 (it is 1/137.035999070). Many famous physicists have tried to understand why its inverse is so close to the integer 137.

In his article "The Fine-Structure Constant: From Eddington's Time to Our Own," Jacob Bekenstein of the Hebrew University says:

> There is a story about this, which I remember from my undergraduate days. Wolfgang Pauli died and went to heaven. God received him and showed him around. "Here is where you'll be staying, Wolfgang. Do you have any questions?" "Yes," said Pauli, "why is [the inverse of the fine structure constant] so close to 137?" "Ah," said God and he handed him a thick packet. "Read this preprint of mine; I explain it all there."[18]

In his book on quantum electrodynamics, *QED: The Strange Theory of Light and Matter*, Richard Feynman says this about the fine structure constant:

> It has been a mystery ever since it was discovered over fifty years ago, and all good theoretical physicists put this number up on their wall and worry about it. Immediately you would like to know where this number for a coupling comes from: is it related to pi, or perhaps to the base of the natural logarithms? Nobody knows. It's one of the *greatest* damn mysteries of physics: a *magic number* that comes to us with no understanding by man. You might say the "hand of God" wrote that number, and "we don't know how he pushed his pencil."[19]

The electromagnetic processes govern the entire field of chemistry, since the interactions of electrons with each other and with nuclei are electromagnetic. But the electron has relatives—particles that are like it in some ways. All of them belong to the same family, called the *leptons*. Leptons can interact through the action of another force—the weak force. We will meet them next.

Chapter 7

"Who Ordered That?"—The Discoveries of Leaping Leptons

The electron was discovered long ago, still in the nineteenth century, as we recall, by J. J. Thomson in 1898. The electron is an *elementary particle*—meaning that it has no internal structure, no components inside it; it is basically like a point, with no volume, and it is endowed with a negative electric charge we define as −1.[1] The electron is a kind of particle now seen to belong to a family called the *leptons,* whose name comes from the Greek word for "thin." Leptons are particles that move freely—unlike quarks, which are confined inside protons, neutrons, or mesons. I like to think of leptons, mnemonically, as particles that can "leap."

The next member of the lepton family to be discovered theoretically was the neutrino. In 1930, Wolfgang Pauli was studying the experimental results of Enrico Fermi on beta decay—a kind of radioactive process in one of whose forms a nucleus of an atom emits an electron. Pauli's calculations told him that in this process there was a missing amount of energy: There was a tiny but measurable difference between the energies and masses of the particles before the decay process and the total mass-energy of all the particles after the decay. Since he knew that energy is conserved in physical processes, Pauli conjectured that an unseen, undetected "new" kind of particle—a very "tiny" one, in the sense that it had very little energy or mass—was also produced in the decay, and that this particle, which had to have a neutral electric charge, accounted for the missing energy in the reaction; now everything would add up nicely and energy would be perfectly conserved.

Pauli chose an unusual way of communicating his theoretical discovery to the world. He presented his prediction in a curious letter addressed to the "Radioactiven Damen und Herren" ("Radioactive Ladies and Gentlemen"), the attendees of the radioactivity session of a physics conference held in December 1930 at the University of Tübingen, in Germany. Pauli's letter, dated December 4, 1930, announced his prediction of the existence of the new particle and urged the conference attendees to search for it in experiments on radiation. Apparently Pauli had decided to write his letter, rather than announce his discovery in a presented paper, because he couldn't bring himself to go to the physics conference, preferring instead to attend a fashionable ball at the Baur au Lac Hotel in Zurich.[2] Two years later, after James Chadwick had experimentally discovered the neutron, a much bigger neutral particle, Enrico Fermi named Pauli's hypothetical particle the *neutrino*—which would be the Italian diminutive of "neutron." The neutrino is thus a "little neutron."

None of Pauli's "Radioactiven Damen und Herren" ever found the theoretical new particle, but in 1956 it was dramatically discovered by the American physicists Clyde Cowan and Frederick Reines in the radiation emanating from a nuclear reactor, the Savannah River Plant in South Carolina. Pauli's bold prediction was brilliantly confirmed.

Today we understand the beta decay reaction as a neutron inside the nucleus turning into a proton, giving rise to the electron and an *antineutrino*. In fact, we now know that this particle interaction consists of a down quark inside a neutron changing to an up quark, which turns the neutron into a proton, releasing the electron and with it the antineutrino.

The electron and the antineutrino are therefore interlinked—they are related to each other because both appear together in Fermi's beta decay process. The force responsible for beta decay is the *weak nuclear force*, or simply the *weak force*, and the particle interaction that happens here, the changing of a neutron into a proton plus the electron and the antineutrino, is called a *weak interaction*. The fact that an antiparticle appears in this reaction demonstrates how antimatter is actually produced in our matter-dominated universe. But the neutron was not discovered until 1932, and quarks were only proposed to exist more than three decades later.

Neutrinos come to us from outer space, and many of them arrive here from the Sun, where they are created in nuclear processes. And as demonstrated by the Cowan and Reines experiment, they are also produced in large numbers in nuclear reactors. The experimental discovery of the neutrino in 1956 caused great excitement in the world of science. Neutrinos can pass through matter, rarely interacting with it because they are tiny and carry no electric charge, and hence do not interact electromagnetically with matter; for example, their passage does not release electrons from atoms they pass on their way.

To find the neutrinos, Cowan and Reines needed to be clever in their experimental design. The two physicists and their collaborators used the very heavy stream of the hypothesized neutrinos believed to be created in a nuclear reactor: *Trillions* of neutrinos per second were believed to pass through every square centimeter of the surface near the source of the radioactivity inside the reactor. The researchers forced these particles to pass through a tank of water, where they thought some of them should interact with protons in the water molecules.

Because neutrinos don't like to associate with other matter, the reactor with its immense rate of neutrino production was necessary so that the rare event of interaction would still occur with a measurable frequency. When this interaction occurs, an antineutrino hits a proton, resulting in the creation of a neutron plus a positron. The positrons, being the antimatter twins of electrons, then annihilate with electrons they meet, producing gamma radiation, whose intensity and direction was analyzed by Cowan and Reines. These measurements, along with an analysis of the resulting neutrons, led to the conclusive discovery of the neutrino.

If you have wondered how the highly sophisticated detectors of the LHC were designed, how scientists had the knowledge of how to detect any kind of particle that results from the proton collisions, knew how to map their tracks, and how to study the directions and energy levels of particles, the answer is that there is much collective experience in this field. The Cowan and Reines experiment of the 1950s demonstrates how experimental physicists think and how they design collisions and detection schemes for even the tiniest and most elusive of particles.

We should note that none of the components of ATLAS and CMS can directly detect neutrinos. The reason for this is that the rate of interaction of neutrinos with other matter is so small that the probability that a detector would reveal such a reaction is minuscule. So for the experiments of the LHC, the existence of neutrinos is inferred by missing energies in particle reactions—just as Pauli had done eighty years ago—rather than through experimental detection. (Cowan and Reines had orders of magnitude more neutrinos than are produced by the LHC and hence could detect them.)

———

As we know, one of the two main general-purpose detectors at the LHC is the *Compact Muon Solenoid* (CMS). The muon is yet another cousin of the electron—that is, a lepton. The muon was discovered in 1936, six years after Pauli's theoretical discovery of the neutrino and twenty years before the neutrino's experimental confirmation by Cowan and Reines. The discovery of the muon was purely accidental—no one had expected the existence of such a particle. The muon is produced naturally in the highest levels of Earth's atmosphere as the result of collisions of cosmic rays—streams of high-speed particles, usually protons, from space—with the nuclei of atoms in the upper atmosphere.

The muon was discovered by Carl D. Anderson, who in 1932 had discovered the positron predicted by Paul Dirac. He was aided in his work by his Caltech graduate student Seth Neddermeyer. Anderson and Neddermeyer identified muons when their tracks in a magnetic field were found to bend in the direction and to the extent that established that they had the same charge as the electron but that their mass was much higher. The design of this experiment is reflected in the LHC, whose detectors all use powerful superconducting magnets to bend particle paths so that they can be studied. Thus the methodology of the LHC detectors has its roots in experiments conducted more than seventy years ago.

The muon weighs as much as about 207 electrons (and about one-ninth as much as a proton). The electron itself weighs several million times more than the neutrino, which has a very tiny mass. Even though it is negatively charged like the electron, the muon is not hindered much by matter and can penetrate it to great depths. An elementary particle, the muon is a *second-generation* lepton, the term physicists use to describe particles that are heavier than the usual matter particles, in this case the electron.

When the muon's discovery was announced in 1936, Isidor Rabi, a leading Columbia University physicist and Nobel Laureate, was dining among a group of physicists in a Chinese restaurant in New York City. When a physicist told him the news, Rabi turned to his colleagues and exclaimed: "Who ordered *that*?"[3] This reaction typified many physicists' surprise at the existence of particles that no one had expected.

The muon is unstable: On average it lives 2 millionths of a second before it decays. When this happens, the muon breaks down into an electron, an antielectron neutrino (this is Pauli's original neutrino from beta decay), and a *muon neutrino*—a new kind of neutrino we will discuss soon. The decay of the muon is governed by the weak force; this is a particle interaction that is thus similar to Fermi's beta decay, which as we know is also a manifestation of the action of the weak force.

The muon lifetime presented a puzzle. If the particle is created in the upper atmosphere, and if its lifetime is so short, how come so many muons are detected at ground level—and even found at some depth underground? The answer was found to be special relativity. For a muon traveling at close to the speed of light, there is *time dilation*: Time intervals become longer, or, equivalently, time ticks slower—as seen from our vantage point on the ground. The slowing down of their clocks is what allows the muons to travel a much greater distance before they disintegrate. The lifetime of a fast-moving muon is one of the best experimental proofs of Einstein's special theory of relativity.

The amazing penetration power of muons was used by Luis Alvarez and his colleagues in a study of the pyramids in the late 1960s.[4] Here, muons produced by cosmic rays were measured to penetrate 300 feet of limestone to finally reach detectors placed in the Belzoni Chamber right under the center of the Second Pyramid at Giza. The muons thus exposed the internal structure of this pyramid, helping scientists solve a long-standing puzzle about the evolution of Pharaonic architecture. The technique involved counting the numbers of muons arriving at the Belzoni Chamber from different angles. The existence of hidden chambers would have shortened the muons' paths through the limestone and hence increased the numbers of muons at certain angles, because with more limestone, more muons would have been absorbed by this dense medium. In the space that was explored, no hidden chambers were found, surprising those scholars who had expected more internal structure in this pyramid.

The study by Alvarez and his colleagues made the first innovative use of muons—which come to us free from nature, as opposed to X-rays, which have to be created at great expense. And muons have properties that X-rays cannot offer, since the latter do not penetrate stone and earth to anywhere near the degree that muons do. More recently, in this century, the muon technique used to look for possible hidden spaces in the Second Pyramid in Giza was suggested as a tool for use by teams of archaeologists searching for potential hollows in Mesoamerican pyramids—the great pyramid at Teotihuacán in Mexico, as well as Mayan pyramids in Belize. And use of this technique is gaining momentum in related fields as well: Muons from cosmic ray impacts are now also used to peer into volcanoes in Japan to determine how serious the danger of an eruption may be.[5]

In 2003, scientists working at the Los Alamos National Laboratory in New Mexico proposed that muons be used as a means for detecting a nuclear device that terrorists might try to bring into the United States in a cargo container.[6] Each minute, 10,000 muons land on every square meter of the surface of the Earth. These muons can be used as natural "X-rays." Nuclear contraband would have to include extremely dense elements, such as uranium, plutonium, and lead, which should be detectable through the use of cosmic muons. Thus the same principle used to look for hidden chambers in pyramids and to reveal the density of matter inside a volcano could also be used effectively to search shipping containers for nuclear material, eliminating the need for the more expensive and time-consuming process of opening the containers.

In 1959, while the physicists at Columbia University were having their Friday afternoon coffee meeting, T. D. Lee, who had won a Nobel Prize two years earlier with C. N. Yang for their joint work on parity violation, posed a question to his colleagues: "How might we study the weak interaction at energies higher than those of particle decays?" A member of the department, Melvin Schwartz, thought about it, and came up with an interesting answer: "Make neutrino beams in particle accelerators, and study them."[7] This suggestion spurred interest in research on neutrinos.

The reason that neutrinos are such a good tool for analyzing the weak interactions, of which Fermi's beta decay is one example, is that almost the only thing that has any effect on neutrinos is the weak force. Neutrinos have no electric charge, so they don't feel any electric attraction or repulsion from other particles and hence are immune to the electromagnetic field. They are not quarks, so gluons can't touch them, which means they are immune to the strong force that acts inside the nucleus. And finally, their mass is so incredibly small that the force of gravity almost doesn't affect them at all. The only force they feel in any measurable way is the weak force.

At the time that Schwartz and two other Columbia physicists, Jack Steinberger and Leon Lederman, made their decision to look at neutrinos, it was becoming experimentally possible to focus precisely the neutrino beams created in accelerators; this technology was being implemented in two new accelerators whose construction had been completed in 1960. One was the Alternating Gradient Synchrotron (AGS) accelerator at the Brookhaven laboratory nearby, and the other was the Proton Synchrotron (PS) machine at CERN.

In 1962, the three scientists performed experiments with neutrinos using the AGS accelerator, but their results were puzzling: The bombardment of metal sheets by protons produced muons and neutrinos, but never any electrons as the physicists had expected would happen. This meant only one thing: The neutrinos that were created with the muons had to be of a different kind from the neutrinos hypothesized by Pauli and discovered by Cowan and Reines.

Pauli's neutrino is now called the "electron neutrino," and the new kind of neutrino found by Lederman, Steinberger, and Schwartz is called the "muon neutrino"—a neutrino associated with the muon, rather than with the electron. This discovery rocked the world of particle physics because it was so unexpected: The elusive, tiny, and almost massless neutrino was not a single particle but had a seemingly identical relative that was still different from it. In 1988, the three physicists were awarded the Nobel Prize for their discovery.

The muon neutrino is a second-generation neutrino, after the electron neutrino. It is defined as a second-generation lepton because it is associated with the muon, which is a second-generation lepton after the electron (and not based on its own mass, which is very tiny and as yet indistinguishable from that of the electron neutrino).

The third neutrino, the tau neutrino, was discovered in 2000 by fifty-four physicists working collaboratively at Fermilab. Leon Lederman had been director of Fermilab until a decade before this discovery. When I asked him about the tau neutrino, he joked: "They should have taken back my Nobel Prize [for his codiscovery of the muon neutrino] because I didn't find the third neutrino!"[8]

The tau neutrino is a third-generation neutrino, associated with a third particle like the electron and the muon: the tau lepton. The very massive tau is a third-generation lepton, discovered in 1975 at the Stanford Linear Accelerator Center (SLAC) by Martin L. Perl, for which he won a Nobel twenty years later.

Martin Perl was born in New York and studied chemical engineering at the Brooklyn Polytechnic Institute, an institution that has recently been absorbed by New York University. Then he worked for a few years for General Electric as a chemical engineer, designing electron tubes. He spent some time in the Merchant Marine, sailing cargo ships along the United States coasts.[9] Eventually he returned to school and through the encouragement of one of his professors went to study physics at Columbia.

"Until I went to Columbia," he told me when I interviewed him at his office at SLAC in Stanford, "I was the smartest person I knew." At Columbia, however, Perl met other bright students and faculty in a physics department that was very stimulating intellectually. And in particular there was a professor with whom Perl wanted to work. "I still don't know how I got the courage to do it," he said, "but I went to Isidor Rabi and asked him if he would take me on as a student—he was famous, had a Nobel Prize."[10] Rabi agreed and soon suggested that Perl go into high-energy physics, which was then a new field with much promise for a young doctoral student.

Having finished his doctorate in physics under Rabi in 1955, Martin Perl got job offers from Yale University, the University of Illinois, and the University of Michigan. He chose Michigan, because he thought that there would be new opportunities at a place that had a developing physics program. Here he worked with Donald Glaser, the inventor of the bubble chamber.

Glaser had invented the bubble chamber in Michigan in 1952 while holding a glass of beer in his hand and wondering how the bubbles formed in it. This puzzle gave him the idea that subatomic particles (nuclei, protons, electrons, muons, and others) passing through a fluid would precipitate the formation of bubbles.[11] He then experimented with different kinds of drinks: beer, soda water, and ginger ale. But none of these liquids enabled him to detect the effects of ionizing radiation, such as the passage of a proton through the liquid in the glass, as he had hoped. Then one day he discovered that ether, when heated to a high temperature, did respond to radiation by creating a trail of bubbles along the path of a particle passing through the liquid—and the bubble chamber was born.

Perl and his colleague Lawrence W. Jones then went to work on a new device in experimental high-energy physics, the spark chamber,

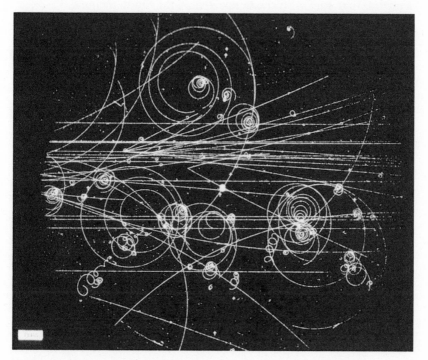

Particle tracks in a small bubble chamber in a magnetic field at CERN

an improved instrument for studying the tracks of particles. Perl received an offer from SLAC, which had just been opened near the Stanford University campus in California, and he moved there in 1963. At SLAC, Martin Perl became interested in muons and did experimental work with these particles.[12]

Then in 1975, Perl made his major discovery. Working at SLAC, he found the third-generation lepton, which he named, with his Greek graduate student Petros Rapidis, the tau, after the Greek letter of that name. The tau lepton weighs almost 3,500 times as much as the electron (its mass, in energy units, is 1,776 MeV, or 1.776 GeV). Like the electron, the tau is an elementary particle with no extent or structure. It's like a point—but it's very heavy. Physicists think that there are only three generations (sometimes called families) of elementary particles, and Perl found the first lepton of the third generation. The third generation of the leptons was completed in 2000 with the discovery of the neutrino that goes with the tau particle—the tau neutrino.

Because it is so heavy, the tau decays in hundreds of possible ways, all of them occurring in a very short interval of time: a trillionth of a second (10^{-12} seconds) after its creation. "No one knows why the masses work the way they do," Perl said to me. "Even if the Higgs is discovered at CERN, we still won't know why the weights are the way they are. The Higgs is supposed to give mass to itself and to other particles, but there is no good explanation as to why, and to how heavy the particles can be. We're just at the beginning of the road on that," he said.[13]

"We understand how matter grows in the periodic table," Perl continued, "but we don't have such an understanding of the elementary particles." We view such particles as points, even if their masses are large. "Maybe information about them is found in the fields around them—is what some people say," he concluded. "We're only at the beginning here—nowhere close to the end of science."[14]

The electron, its heavier relative the muon, and the third and much heavier "cousin," the tau, along with their respective neutrinos, complete the lepton group of elementary particles. It is shown below.

THE LEPTONS		
First Generation	*Second Generation*	*Third Generation*
electron	muon	tau
electron neutrino	muon neutrino	tau neutrino

One can create a corresponding table, which would have the prefix "anti" attached to every particle. Such a table would represent the antiparticles: antielectron (positron), antimuon, antitau, anti–electron neutrino, anti–muon neutrino, and anti–tau neutrino.

There are ongoing studies of neutrino *oscillations* at CERN and other research facilities around the world. In an *oscillation,* a neutrino of one kind can turn into a neutrino of another kind—for example, an electron neutrino may suddenly turn into a muon neutrino. This process is somewhat mysterious within our present knowledge of physics, but it has one important implication: Such an oscillation implies that the oscillating particle has mass. And ever since this discovery was first made at the Super-Kamiokande neutrino observatory in Japan in 1998, we have known that neutrinos have mass.

The group of particles called the leptons forms one sector of the Standard Model. Another sector is that of the quarks, and a third one is that of the force-carrying particles, which are bosons. We will encounter both new groups of particles in chapter 8, allowing us to construct the full table of the Standard Model of particle physics. All these particles play crucial roles in the interactions that take place inside CERN's Large Hadron Collider.

Chapter 8

Symmetries of Nature, Yang-Mills Theory, and Quarks

Ever since the discovery of the neutron in 1932, physicists have been intrigued by the nature of the twin types of particles that live inside the nucleus: the proton and the neutron. The electron, which normally circles around the nucleus, is about 1,800 times lighter than the proton and neutron. But the proton and the neutron have *almost the same mass*. This fact presented an intriguing puzzle: *Why are the masses of these two particles almost identical?* And what does this say about the nature of these particles? Are they related to each other in some fundamental way? Could they, perhaps, be *two different manifestations of the same particle*?

Right after the discovery of the neutron, Werner Heisenberg tried to address this mystery by hypothesizing that an abstract mathematical *symmetry* of nature linked the two particles to each other. And since quantum mechanics is ruled by the idea of the superposition of the waves associated with particles (these waves are the solutions of Schrödinger's equation), a symmetry between two particles implies that *one can be continuously "rotated" into the other.* This "rotation" takes place in an abstract mathematical space—not a physical one. The rotation, or *deformation,* of one particle into another is the continuous mixing of one with the other according to quantum rules. For example, instead of a proton or a neutron, at any given moment in time we might have a *mixture* of these two *states*: "being a proton," and "being a neutron." We may thus have something that is, say, 23 percent a proton and 77 percent a neutron, or a composite entity that is 39 percent a proton and 61 percent a neutron, and so on. This is just the way the quantum world works, because particles are also waves, and waves—by their very nature—can be superimposed on one another in this way. The quantum world thus allows for a continuous rotation, or deformation, of one particle into another, when both are viewed as merely two different states of the same entity. At any given moment in time what we have is not purely a particle of one kind, but rather some mixture of the two kinds of particles put together.

This is a difficult concept to accept or understand, because it doesn't correspond to any intuition we might have about the world from our everyday experience; but the quantum realm follows its own bizarre laws. Think of Schrödinger's Cat paradigm: The cat is *both dead and alive at the same time*. Put another way, Schrödinger's Cat can be continuously rotated from the state of being alive to the state of being dead and vice versa. The same quantum logic implies that a proton could somehow become a neutron by being continuously "rotated" into that state and vice versa.

The symmetry that Heisenberg proposed for linking the proton with the neutron is called *isospin* symmetry, or isotopic spin symmetry. The isospin is a *quantum number* whose value differentiates between the proton and the neutron. The proton is defined to have an isospin equal to ½, while the neutron has isospin equal to -½.

This idea is a direct extension of the idea of spin. Recall that the spin of the electron is defined to be ½. But because of quantum behavior, the actual spin at any given moment is a superposition of two spins: +½ and -½. The same idea works for isospin: The proton's isospin is ½ and the neutron's is -½, so we can have a superposition of a proton and a neutron, which is something whose actual isospin at any given moment is a combination of these two numbers. The superposition can

Proton   Neutron

consist of, for example, 30 percent isospin ½ and 70 percent isospin –½. The superposition of the proton and the neutron is shown above.

Heisenberg's idea of a symmetry linking two particles that look similar was taken up two decades later by the Chinese American physicist C. N. Yang, whom we have encountered earlier when we discussed his joint work with T. D. Lee on parity violations. Earlier, Yang had done important work on symmetry together with his colleague Robert L. Mills at Brookhaven National Laboratory in 1954. Their results completely changed theoretical physics, and their implications still reverberate not only throughout the field of physics but also in the realm of pure mathematics.

The Yang-Mills theory underlies the theoretical foundation of much that we may learn from the experiments at the LHC. This theory forms a framework for physical models that are called *gauge theories,* and they are among the most powerful theoretical tools in modern physics. A gauge theory is one that connects the symmetry in a physical situation with the properties of some continuous group, such as the circle group we have seen earlier. (Recall that such groups are called Lie groups.) The term "gauge" had been invented in the 1930s by the German mathematician Hermann Weyl.

Here is an example. You can rotate the wedding band on your finger without anyone noticing any difference in how your ring looks once you have rotated it. We could say that the symmetry of the rotations of your wedding band is *gauged* to the continuous group of rotations of the circle. Now suppose that you visit a fortune teller with a crystal ball.

She picks up the ball, rotates it in space (any rotation is a combination of two basic spatial rotations, which we can take conveniently to be along a "latitude" and along a "longitude," as on a globe), and puts it down again. If the ball has no markings on it, you will not be able to tell that it had been rotated. The symmetry here (the fact that you can't tell what was done to the crystal ball) is *gauged* to a continuous group of the rotations of a sphere. In physics, often such rotations and other transformations take place in an abstract mathematical space, rather than a physical one—but the idea is exactly the same.

Chen Ning Yang (known as C. N. Yang) was born in Heifei, in the Anhui Province of China, in 1922. His father, K. C. Yang, received a PhD in mathematics from the University of Chicago and then returned to China to teach mathematics at the Tsinghua University in Beijing and later at the National Southwest Associated University in Kunming. Yang's father saw that his son was interested in algebra as a young student and gave him books about this topic and encouraged his curiosity. C. N. Yang became fascinated with the ideas of symmetry and group theory and how the structures of groups could be used to reveal the recondite symmetries of nature, such as the possible rotations of crystals and rotations in more abstract settings.

Yang graduated in 1942 and continued on to study at the graduate level, taking various courses in physics and learning about field theory as well as statistical mechanics. But in the mathematical theory of symmetry he saw the key to unveiling the mysteries of nature. Yang came to the United States after World War II and went on to earn a doctorate in physics from his father's alma mater, the University of Chicago.

After finishing his doctorate, Yang spent the year 1953–1954 at the Brookhaven National Laboratory. Here he met Robert L. Mills, and the two of them developed what is now known as the Yang-Mills theory. The Yang-Mills theory is based on groups and symmetries and is an extension of the theory of the electromagnetic interactions. Yang and Mills succeeded in modeling the isospin symmetry between the proton and the neutron. This was a very difficult task, which had stumped many physicists before them, because the isospin symmetry is *non-Abelian.* "Abelian" refers to Niels Henrik Abel, a nineteenth-century Norwegian mathematician who was a contemporary of Galois and studied the same ideas of symmetry and groups that Galois did. Like Galois, Abel died young: from consumption, at age twenty-seven. Abelian groups, named after him, are groups in which the group operation is *commutative,* meaning that the order of the operations does not make a difference to the final answer. The operation of putting on your shirt and putting on your pants are commutative, for example: It doesn't matter which you do first—at the end you'll be dressed. But the two operations of putting on your pants and putting on your underwear are noncommutative: It makes a big difference to the final outcome which of these you do first.

The quantum electrodynamics theory of Feynman, Schwinger, and Tomonaga is Abelian—that is, commutative. The group involved is mathematically easier to handle than non-Abelian (that is, noncommutative) groups. Things become much more complicated in the case of the proton together with the neutron. Here, as often happens in quantum mechanics, the basic operations describing particle interactions are badly noncommutative.

Quantum mechanics as a whole is a distinctly noncommutative theory. In quantum mechanics, when you measure one variable, it immediately affects your precision in measuring another variable (this is Heisenberg's uncertainty principle in action); therefore the *order* of measurements is important. Measuring the position of a particle and then measuring its velocity will not yield the same result as measuring the velocity first and then measuring the position, because in each case you will know the first quantity you measure with good precision and the second with less precision. We say that these two measurement operations do not commute. Therefore, they must be handled by mathematics that is noncommutative. The noncommutativity requirement of quantum mechanics and quantum field theory hindered progress in studies of symmetry in theoretical physics—until the groundbreaking work of Yang and Mills. To model Heisenberg's isospin symmetry, Yang and Mills used a different Lie group than that of the circle. They used a special continuous group that allows many more kinds of rotations in an abstract space.[1] This group continuously mixes two kinds of entities: a proton and a neutron.

Yang-Mills theory was then applied very successfully in a wide variety of theoretical situations. It underlies the very structure of the Standard Model of particle physics; in fact, theoretical considerations have shown in various contexts in physics that often the only kind of theory that works well is a Yang-Mills gauge theory. This theory aided the theoretical discovery of quarks—the up and down quarks that make up protons and neutrons, as well as the two upper generations of quarks, which are heavier and unstable. The Lie group used here was of a higher order: It was the group that continuously mixes *three* entities—the hidden quarks, three of which live inside every proton and neutron—rather than two entities, as in the isospin mixing of the proton and the neutron.

Murray Gell-Mann, the outspoken theoretical codiscoverer of quarks, was born in New York in 1929. He was accepted to Yale University at age fifteen, and went on to obtain his doctorate in physics from MIT.

In 1964 Gell-Mann hypothesized the existence of quarks, a proposal also made independently that same year by George Zweig, then a graduate student at Caltech. Gell-Mann chose to describe them by the word "quarks," which he got from James Joyce's novel *Finnegan's Wake* (where a quark means a screech); Zweig named them "aces." Another name was given to these constituents of neutrons and protons by Richard Feynman. He called them "partons"—because they constitute "parts" of the nucleons; some people still use the term "partons" to refer to quarks (together with gluons).

Frank Wilczek, who in 1973 was to play a crucial role in explaining the behavior of quarks through a theory he codeveloped with David Gross, independently formulated by Hugh David Politzer (the three men shared the 2004 Nobel Prize), tells a funny story about partons in his book *The Lightness of Being*. He says that Murray Gell-Mann was infuriated every time he heard anyone call the putative constituents of hadrons "partons," as Feynman referred to them.

Wilczek recalled that he once told Gell-Mann, "I'm trying to improve the parton model," and added: ". . . so here I confess that it wasn't entirely in innocence and ignorance that I mentioned partons. I was curious to see how Gell-Mann would react to his rival's idiom."[2] Wilczek apparently got what he expected:

> Gell-Mann gave me a quizzical look. "Partons?" Stage pause, facial expression of deep concentration. "Partons?? What are partons?" Then he paused again and looked very thoughtful, until suddenly his face brightened. "Oh, you must mean those *put-ons* that Dick Feynman talks about! The particles that don't obey quantum field theory. There's no such thing. They're just quarks. You shouldn't let Feynman pollute the language of science with his jokes."[3]

Incidentally, Wilczek, Gross, and Politzer would later explain how partons, or quarks, do obey a quantum field theory—the theory of quantum chromodynamics developed by these three physicists.

In the 1960s, Gell-Mann worked on classifying the hadrons—the group of particles that includes the nucleons and mesons. Today, thanks to Gell-Mann and, independently, to Zweig, we know that these are not elementary particles but rather are composed of quarks: three quarks in a nucleon (and other baryons) and a quark-antiquark pair in the case of the mesons (they don't annihilate each other because they are of different kinds, or a quantum mixture). Quarks were hypothesized to exist because of a stream of discoveries of new particles now understood to be composed of quarks. These curious new unstable particles shed light on the constitution of the ordinary matter of everyday life. We now know that all the usual, stable matter in the universe is made up of a surprisingly small number of elementary particles: only three! At its most fundamental level, the universe of the things we see around us is made up of the three elementary particles:

The up quark
The down quark
The electron

The two kinds of quarks make up protons and neutrons, which live in the nuclei of matter. Each proton contains two up quarks and a down quark, and each neutron is made of two down quarks and an up quark. The electrons orbit the nucleus, completing the atom. But other kinds of quarks—in addition to the two kinds (up and down) that live inside the proton and the neutron—were found as well.

When a new meson was discovered in 1947, which decayed much slower than physicists had expected, it was called a *strange* particle. So when quarks were postulated by Gell-Mann and Zweig some decades later, the strange meson was understood to contain a *third* type of quark, in addition to the up and down quarks that make up protons and neutrons. It is called the strange quark.

Later, in the 1970s, a "charm" quark, and a "beauty" (or "bottom") quark were discovered—names were fanciful—and the table of elementary particles of the Standard Model was getting filled up. The existence of the charm quark was inferred from the discovery in 1974 of a meson that contained it by Samuel Ting, working at Brookhaven, and Burton Richter, working at SLAC, who shared the Nobel Prize two years later. The beauty quark was discovered in 1977 through a large experiment at Fermilab. On the lepton part of the table, the tau was added; and there are three kinds of neutrinos, as we have seen. Finally, the discovery of the very heavy *top quark* was reported in 1995 by two large international collaborations of scientists working at Fermilab. This brought the number of known quarks to six (and of course there are six anti-quarks as well), and it completed the fermions for the Standard Model.

The quarks have *fractional* charges, as compared with the charge of the electron, defined as −1. The charge of the up quark is ⅔, and that of the down quark is −⅓. Let's see how this works. The proton is made of two up quarks and a down quark, so its charge is equal to ⅔ + ⅔ + (−⅓) = 1, as it should be (the proton's charge is exactly the opposite of the electron's charge). The neutron is made of two down quarks and an up quark, so its charge is: −⅓ + (−⅓) + ⅔ = 0, which checks out as well, since we know that the neutron is neutral—it has no charge (its charge is zero). The same pattern holds for the next two generations of quarks: The charm quark has charge ⅔ and the strange quark −⅓; and the top quark has charge ⅔, while the bottom quark has charge −⅓. The quantity and sign of the charges thus follow the exact pattern exhibited by the up and down quarks of ordinary matter. For antiparticles, all the signs of the particles above are reversed, giving a total charge of −1 for the antiproton, as it should be, and the antineutron has zero charge.

The masses of the three generations of quarks vary wildly. The mass of the up quark is about 4 times that of the electron (whose mass is about 0.5 MeV), while the top quark, the heaviest of them all, weighs roughly 178 GeV, which is about 90,000 times the mass of the up quark, around 360,000 times the mass of the electron, and about the same mass as that of a gold nucleus.[4] The down quark is slightly heavier than the up quark, and hence the neutron, which contains an up quark and two down quarks, is slightly heavier than the proton, which contains two up quarks and a down quark. The strange quark weighs about 0.1 GeV—that is, about 200 electron masses; the charm quark weighs in at about 1.27 GeV, more than 2,000 electrons; and the bottom quark has a mass of around 4.2 GeV, or as much as 8,000 electrons. Some of these numbers have been predicted by theoretical physicists.

Mary K. Gaillard was born in New Brunswick, New Jersey, in 1939; graduated from Hollins College in Virginia in 1960; and received a master's degree in physics from Columbia in 1961. She went on to earn two doctorates in science from the University of Paris in Orsay and later took a position with the French National Center for Scientific Research (the Centre national de la recherche scientifique, or CNRS). While employed by this French agency, she worked at CERN, where she collaborated on the famous "penguin paper" with John Ellis and their coauthors.

Gaillard also worked at Fermilab, and in 1981 she became the first woman professor of physics at the University of California at Berkeley. In 1973, together with Ben W. Lee, she predicted the mass of the charm quark, before it was discovered. Her joint prediction was of a mass of about 1.5 GeV. In studies conducted since the particle was found, its mass has been estimated as 1.27 GeV. Leon Lederman's group had predicted that the quark's mass would be around 3 GeV, but Gaillard and Lee's estimate was much closer.[5]

Later, together with M. S. Chanowitz and John Ellis, Gaillard predicted the mass of the bottom quark. When they wrote their paper, predicting a particle mass "2 to 5" times the mass of the tau lepton (which weighs 1.776 GeV), John's writing was misread by the editor, who thought the handwritten "to" was the number 60, so in the abstract the predicted mass appeared as 2605 times the mass of the tau—a fantastically huge mass. This error was later corrected. The predicted value was found close to the present estimate of around 4.2 GeV.[6]

Quarks are *confined*: A quark alone, all by itself, can never be seen in nature. While the leptons can "leap," quarks must stay put. The reason for this is that they are held together by an immense and very strange force: the strong force. The strong force is very different from the forces we know from everyday life: gravity and electromagnetism. These two forces *weaken* with distance, while the strong force becomes *stronger* with distance.

It is as if a rubber band—a very, very powerful rubber band—was at play here. When the rubber band is relaxed, the quarks can move about within the space enclosed by the band. But if they try to move farther away from each other, they stretch the rubber band and its increased tension brings them closer together. The farther they try to stretch the band, the greater the force the band exerts on them to return.

It's a bizarre phenomenon, but it was proved mathematically by the work of David Gross, Frank Wilczek, and Hugh David Politzer a decade after Gell-Mann's work. Their research showed theoretically that individual quarks can *never* be seen.

In 2004, Gross, Wilczek, and Politzer jointly received the Nobel Prize in physics for their 1973 work on explaining quark confinement. Their theory is called quantum chromodynamics, where "chromo" comes from the Greek word "chroma," which means "color." The strong force acts on the quarks through the massless bosons called gluons, which are similar to the photon but carry a charge called "color"— although it has nothing to do with real color; the term is simply a designation. The symmetry here is a continuous threefold one, because the quarks and the gluons exchange among themselves *three* "color charges."[7]

When protons crash at immense energies inside the LHC, many particles with all kinds of energy levels and masses are produced. The leptons are detected singly, and so are the bosons and other particles. But a quark inside a detector of the LHC is seen through a *jet*—a stream of particles still bound together by the strong force; it sprays out of the collision area. The number of jets that scientists see provides them with information about what else is being produced in the collisions.

In the early universe, at temperatures in the trillions of degrees Celsius, the quarks floated around in a quark-gluon plasma; as mentioned earlier, this is a kind of very high-energy fluid in which some of the particles are charged. This strange state of matter is the focus of the ALICE collaboration at the LHC.

Quarks, Gluons, and the ALICE Experiment

As noted earlier, about 10 percent of the time that the LHC is running, it is not smashing the usual protons, but rather the very heavy nuclei of lead. The purpose of this experiment is to reproduce the quark-gluon plasma, sometimes called "quark soup," which permeated the universe after the Big Bang. According to theory, the quarks and gluons formed this plasma when the temperature was about 2 trillion degrees Celsius, which is more than a hundred thousand times hotter than the center of the Sun. For a few millionths of a second after the Big Bang, temperatures are believed to have been that high, before matter became baryonic as the baryons, the protons and the neutrons, appeared.

Physicists in the ALICE collaboration hope to repeatedly create the quark-gluon plasma for what will have to be extremely short periods of time inside a space that is about the size of an atomic nucleus. Physicists expect that studies of the plasma will lead to information about the actual process of quark confinement, which so far is understood only through mathematical models. They also want to probe the nature of the vacuum of space, which may yield some of its secrets as the plasma suddenly forms and then quickly disappears. They want to learn about the generation of massive particles under the strong force carried by the gluons. They want to know why protons and neutrons weigh a hundred times more than the quarks they contain. And they want to test the limits of the theoretically derived process of confinement itself: Can confinement be broken under some circumstances?

How does a plasma behave? While we don't yet know how the primordial soup of quarks and gluons that inhabited space after the Big Bang behaved, we do know a lot about other forms of matter in which large numbers of particles float. One such fluid is called a Bose-Einstein condensate. In this special case of a gaseous collection of particles, scientists see the laws of quantum mechanics working in a special, eerie way. Experiments in this direction have been carried out at very cold—rather than intensely hot—conditions.

We know that the principles of quantum mechanics apply to any very small system. Wave properties such as diffraction and interference are apparent in the behavior of small particles of every kind: photons, electrons, protons, neutrons, nuclei, and atoms. These behavior patterns have also been seen with molecules—even some that are so large they contain 60 atoms. Such molecules, named Bucky balls after Buckminster Fuller, whose dome structure they reproduce on a microscale, have been found to exhibit wave properties by a team led by Anton Zeilinger of the University of Vienna.[8] But somewhere, as things grow in size, the wavelike phenomenon disappears. In the macroworld in which we live, "quantum magic" doesn't work. An exception is a Bose-Einstein condensate.

In 1925, Einstein predicted the phenomenon we now call Bose-Einstein condensation based on a method introduced by the Indian physicist Satyendra Nath Bose, after whom bosons were named. When a gas of atoms that are bosons—meaning that they have an integer spin, rather than the half-integer spin of fermions—is cooled to near absolute zero, a large fraction of the atoms condenses in the lowest quantum state possible for them. When the atoms are cooled to the point in which the wave function of each atom has length about the size of the separation distance between the atoms, the wave functions of the atoms *overlap,* and the gas becomes a uniform macrosystem. It is now a kind of quantum soup, made of indistinguishable particles.[9]

In 1995, Eric Cornell and Carl Wieman of the University of Colorado at Boulder and the National Institute of Standards and Technology used rubidium atoms to create a Bose-Einstein condensate. Shortly afterward, Wolfgang Ketterle of MIT and his team achieved such a state using a gas of sodium atoms in their lab. When the temperature was cooled to a very low level, suddenly the Bose-Einstein condensate appeared: The atoms were transformed into one giant "matter wave." This was a very dramatic transition, which Wolfgang Ketterle has described to me this way: "Imagine that you live on a hot planet on which nobody has ever seen ice or snow. You work hard and design and build a refrigerator. Then you place some water in it, and sometime later you open the door and suddenly find a strange and beautiful new form of matter that no one has ever seen before: *ice*!"[10]

When the same kind of cooling that was done to create a Bose-Einstein condensate is applied to fermions, we reach a different kind of state. What we create then is a "Fermi sea."[11] Fermions must obey Pauli's exclusion principle, which means that no two of them can occupy the same quantum state. We saw the exclusion principle in a limited sense in chapter 2, where it was noted that two electrons in the same orbit must have opposite spins.

Pauli's exclusion principle is like a rule a company might have for the parking of its executives' cars: No two BMWs of the same model and color may be parked in the same section of the parking lot. If you have a blue 5-series BMW, you may not park it in Area "A" if another executive has already parked his blue 5-series BMW there. If you really must park there, then paint your car red or any color other than blue. Or change to another car model. Fermions, particles with half-integer spins (such as ½ or 1½), which may be elementary particles or composite ones, must obey this law: If they want to be in the same place, they must differ on at least one other characteristic.

Deborah S. Jin at Boulder; Rudi Grimm in Innsbruck, Austria; and Wolfgang Ketterle at MIT have independently studied the behavior of fermion gases. Such gases are made up of fermions; they have half-integer spins, so they must obey Pauli's exclusion principle. In Ketterle's experiment, a fermion gas was cooled to a very low temperature. The fermions then were seen to form *pairs,* in which each member of a pair had a spin that was opposite of the spin of its partner—as required by Pauli's exclusion principle. The pairs themselves acted like bosons, as the sum of the spins of the two members of each pair was now an integer. Then an experiment was carried out with the population of atoms no longer matched in pairs: spin up, spin down. What happened was fascinating.

Some of the atoms paired up so that one member of each pair had spin up and the other had its spin down—just as Pauli would have dictated. These pairs of atoms lived in a central cloud inside the supercooled chamber. But to the *outside* of this cloud were banished the atoms that could not find partners. They couldn't stay in the exclusive group of pairs in the center, so they just hung around at the periphery. Atoms, in this way, behaved just like unmatched men and women at a dance, standing around looking for partners—as Ketterle has described it.[12] We will soon see the importance of these ideas.

———

So how do we know that the hypothetical particles called quarks—believed to be forever confined so no one can see them by themselves—actually exist? The quarks, while they can never be seen outside their confining baryons or mesons, have been detected in experiments by a team headed by Jerome Friedman and Henry Kendall of MIT and Richard Taylor of SLAC.

Jerome I. Friedman was born in Chicago in 1930 to parents who had immigrated to the United States from Russia in the second decade of the twentieth century. His mother worked in the garment industry, and his father worked for the Singer Sewing Machine Company. Some years after Friedman's parents were married, his father opened a small business repairing and selling sewing machines. Life was hard for new immigrants, but the Friedmans impressed on their children the love of learning, reading, and striving to obtain an education despite the fact that they had little education themselves. There were many books at home, and the parents encouraged their children to pursue art and music in addition to academics. Young Jerome had a gift for art and was leaning toward becoming an art major in college. But while still in high school and very much engrossed in the art program, he read Einstein's book *The Theory of Relativity* and was utterly captivated.

Friedman had an offer of a scholarship to the School of the Art

Institute of Chicago, but he surprised everyone by electing to study physics instead. He applied to the University of Chicago because it had a great physics program that was led by the legendary Enrico Fermi. Having declined an art scholarship, he was now awarded a full scholarship to study at Chicago, without which his family would not have been able to afford his education.

After he had completed the requirements to begin doctoral research, Friedman gathered all his courage and asked Fermi to be his adviser. To his astonishment, the renowned Italian physicist responded favorably. "I felt as though I had won the lottery," Friedman recalled.[13] "Fermi was so smart, it was scary. He was an absolute genius, but he avoided embarrassing his colleagues and students with his brilliance."

Fermi had performed scattering experiments to study the forces affecting subatomic particles, and Friedman used this technique in his doctoral work. But Fermi died of cancer before he could sign off on Friedman's dissertation. This posed a big problem: Others on the faculty would sign on Friedman as a doctoral candidate only if he worked on projects that were of interest to them. Finally John Marshall, another member of the faculty, saved Friedman when he said: "Just go ahead and write it up; I'll sign it."[14]

The recently constructed cyclotron at the University of Chicago was the site of much exciting work at that time. Experiments were conducted there by colliding protons on targets, producing pions (also called pi-mesons), and then scattering these pions off various other targets, such as vessels filled with liquid hydrogen. Friedman used the cyclotron in studying the results of elastic and inelastic scattering of particles. (As discussed earlier, inelastic collisions are those in which energy is transferred to other particles, transforming them or producing additional particles, whereas elastic scattering is like the collision of billiard balls.)

Friedman then accepted a position at the high-energy physics laboratory of Stanford University, which had a small linear accelerator at that time. There he learned the techniques of electron scattering and performed experiments bouncing electrons off nuclei of atoms. He worked in the group headed by Robert Hofstadter, and one of Friedman's associates was Henry Kendall. He also met Richard Taylor, who worked with another group. During this time Friedman developed new techniques of electron scattering, which would prove invaluable in his future work.

In 1960 Friedman took a position in the physics department at MIT. At that time, Wolfgang Panofsky was building the Stanford Linear Accelerator Center (SLAC), which had an energy of 20 GeV. Friedman began to work there, commuting across the country from MIT. At SLAC, Friedman, Kendall, and Taylor formed a new research group and were able to perform precision experiments of scattering electrons inelastically from protons and neutrons to try to "look inside" the nucleons to see if they had any internal structure.

Between 1967 and 1975, Friedman and his MIT and SLAC colleagues found the first direct evidence for the existence of quarks. They used the SLAC accelerator to perform a kind of "electron microscope" experiment to view the interiors of the nucleons. They found three quarks in each nucleon, with the fractional electric charge predicted by the quark theory of Gell-Mann and Zweig. "We weren't looking for quarks," Friedman told me. "We were simply peering into the proton and neutron using inelastic electron scattering at these higher energies. No one was looking for quarks."[15] The physicists were stunned by what they had found, and then they continued to measure the characteristics of the constituents they had observed inside the proton and neutron.

By 1972, the picture became clear: quarks were real! "It was a team effort," Friedman stressed. "Everyone made important contributions. Making the discovery was enabled by our looking at the problem in the right way." By comparing the results of scattering electrons with similar results from scattering neutrinos (which have no electric charge) off protons and neutrons at the Gargamelle bubble chamber at CERN, the fractional charges of the constituents of the protons and neutrons were found to be the same as those of the proposed quarks.

Neutron
Two down quarks and an up quark

Proton
Two up
quarks and a
down quark

Since the neutrino has zero electric charge, its angle of deflection by a quark is independent of the charge of the quark, but the deflection of an electron is affected by its electric charge. So the neutrino scattering gave a comparison against which to gauge the scattering of electrons and thus provided a good basis for determining the charges of the quarks; these agreed with theory. This was a very ingenious method.

The higher energies provided by SLAC allowed researchers to probe smaller and smaller distances by shortening the wavelengths of the electrons. Remember that a particle is also a wave. Increased energy means increased frequency—more wave oscillations per time period, and hence a smaller wavelength. The electrons were able to probe the insides of the nucleons in a similar way to how X-rays—photons of shorter wavelength and higher frequency than visible light—can expose the internal structure of the human body. But the effective magnification power of the SLAC "microscope" peering into the proton and neutron was billions of times more than the power of an ordinary optical microscope.

"Think of the quarks moving very rapidly inside the proton," Friedman said to me. We were meeting over coffee and croissants in a European-style café in Boston on a sunny but freezing December afternoon after he had returned from CERN, where he and other Nobel Laureates had talked about their work while low-energy collisions were taking place in the LHC at the end of 2009. "If you wanted to take a photograph of a very fast-moving object, what would you do?" he asked with a smile. "Use a very fast shutter?" I said. "Yes!" he exclaimed, "But remember, there is a quantum principle here," referring to Heisenberg's uncertainty principle, which says that if a time interval is very small, the energy exchanged in inelastic scattering must be very large. "So we knew that if we got the right amount of energy, we could see the constituents of the nucleons—and we got 2 to 4 billion electron volts," Friedman explained.[16] That was 2 to 4 GeV, enough to capture images of the quarks, seen as points inside a proton or a neutron.

"My goal in my career has been both to understand nature and to contribute to the field," Friedman said, "and I'm very happy that I've been able to do both to some extent."[17] The work by Friedman and his colleagues provided the experimental proof of the existence of quarks—nature's perpetual prisoners.

When the universe was very young, a fraction of a second old—a period studied at the LHC—the quarks lived in a larger enclosure, the quark-gluon plasma (one view is that they were not confined, because they could move freely within the plasma). Here, they were still unable to escape the gluons, but because the energy was so high, the quarks and gluons were free to move in this medium. Only later, when the temperature cooled, did the quarks bunch into threes and get confined in the small units of protons and neutrons. They can also appear in short-lived mesons, which are unstable quark-antiquark pairs. Contrary to the saying "Two's company, three's a crowd," the rule in particle physics seems to be: "Three's company, two's not enough."

The protons and the neutrons (the nucleons) live inside the nucleus of an atom. There, they are held together by the strong nuclear force, which also holds the quarks together inside a nucleon. The residual force from the strong force that holds the quarks inside the nucleon—what "remains" from this extremely powerful confinement of the quarks—is what holds the nucleons together inside the nucleus of the atom. The nucleons interact with the strong force through

The ATLAS detector under construction inside the LHC cavity, showing its characteristic toroid shape and the eight giant rings of its superconducting magnet.

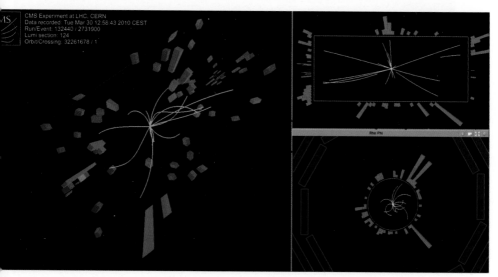

The results of the very first high-energy proton collisions inside the CMS detector at 12:58 p.m. on March 30, 2010.

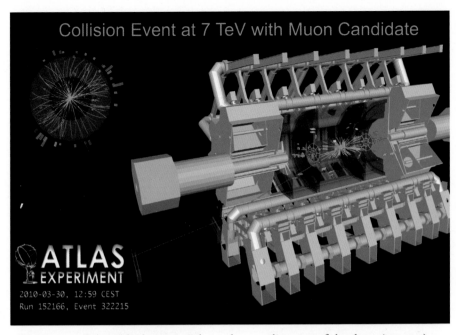

A muon candidate (the long particle track extending out of the detection area) resulting from the high-energy collisions inside the ATLAS detector on March 30, 2010.

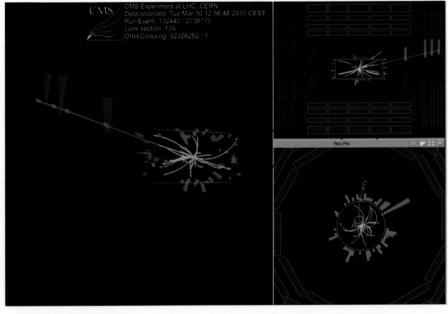

Particle collisions at 7 TeV from the CMS detector on March 30, 2010, showing a muon candidate.

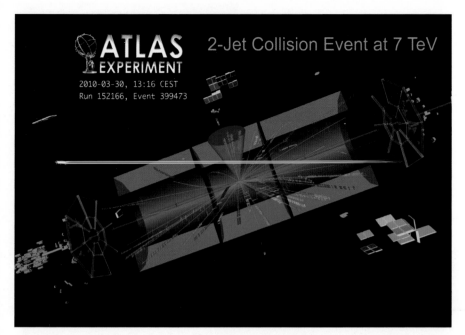

ATLAS collision tracks at 7 TeV, showing two jets (the particle tracks inside the two red cones) emanating from the proton collision point.

The CMS detector before the LHC tunnel was closed for the start of proton collisions.

Leonard Susskind contemplating black holes by the ruins of an ancient Greek temple in Sicily.

Stefano Redaelli at the CERN Control Center on March 5, 2010.

Nobel Laureate Steven Weinberg of the University of Texas, who developed a theory unifying the weak and electromagnetic forces.

Gabriele Veneziano, the inventor of str theory, in his office at the Collège de France in Paris.

Nobel Laureate Leon Lederman of the University of Chicago and Fermilab, co-discoverer of the muon neutrino.

Italian physicist Paolo Petagna inside the cavity of the CMS detector.

Particle physicist Manuela Cirilli at the ATLAS Control Center.

A jubilant Guido Tonelli, head of the CMS group, pointing to the first results of proton collisions at the record energy of 7 TeV obtained by his group on March 30, 2010.

Nobel Laureate Frank Wilczek of MIT, who co-developed a theory of the confinement of quarks.

Fabiola Gianotti, head of the ATLAS group, inside the ATLAS detector during constuction.

Nobel Laureaete Jerome Friedman of MIT, whose experimental work at SLAC provided evidence for the existence of quarks.

Nobel Laureate Martin Perl of SLAC, discoverer of the tau lepton.

Nobel Laureate Philip Anderson of Princeton, whose theoretical work paved the way to our understanding of the Higgs mechanism.

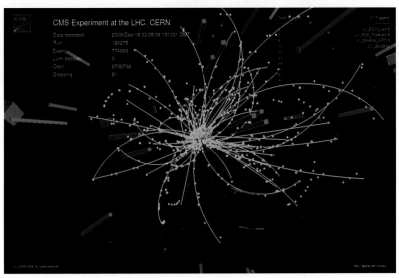

An early CMS proton collision display from December 2009.

An aerial photograph of the border region of Switzerland and France near Geneva, with the Alps in the background, showing the track of the LHC and the locations of its detectors, with inset diagrams of ATLAS and CMS.

a mechanism discovered by the Japanese physicist Hideki Yukawa in the 1930s, which involves the exchange of pions. Thus the nucleus is held together, and the electrons orbit around the nucleus, completing the picture of the entire atom.

Particles made of heavier quarks—charm, strange, top, and bottom—are ephemeral. They are not constituents of ordinary matter but are high-energy particles that appear from collisions of cosmic rays with nuclei in the upper atmosphere or are created in accelerators; in either case, they live very short lifetimes.

THE QUARKS		
First Generation	*Second Generation*	*Third Generation*
up quark	charm quark	top quark
down quark	strange quark	bottom (beauty) quark

Viewed together, the two families of particles, the quarks and the leptons (together they are the fermions), along with the particles that carry force, the bosons, make up the *Standard Model* of particle physics. The bosons consist of the photon and the gluons, which we have met, and also the W and the Z bosons, which mediate the action of the weak force. The Standard Model with all its particles is shown below (and there is a similar table with "anti" as a prefix to every particle's name).

THE STANDARD MODEL OF PARTICLE PHYSICS				
	Fermions			Bosons
Generation: *I*		*II*	*III*	
QUARKS:	up	charm	top	photon
	down	strange	bottom	gluon
LEPTONS:	electron	muon	tau	Z^0
	electron-	muon-	tau-	W^+
	neutrino	neutrino	neutrino	W^-
HYPOTHESIZED SCALAR BOSON:[18]				The Higgs

The Standard Model reflects symmetries that existed in the universe right after the Big Bang, some of which got spontaneously broken, and whose remnants we can see today. The unified symmetry of the Standard Model is captured by the *product* of three Lie groups. These are the group of the continuous rotations of the circle (which models electromagnetism); the group that continuously rotates two kinds of entities into one another (an up quark and a down quark, or an electron and its neutrino); and the group that continuously mixes three kinds of particles (the three quarks in a nucleon, each with a different "color").[19] We will learn more about how this model was put together in later chapters.

The very last missing piece of the entire Standard Model of particle physics is the as yet undiscovered Higgs boson. This is the particle believed responsible for endowing massive particles, including itself, with mass. The existence of the Higgs is implied by the discovery of the W and Z bosons in one of the greatest triumphs of the Standard Model and of modern physics in general. The theory of Glashow, Weinberg, and Salam, which predicted the existence of the three bosons, W^+, W^- (W stands for "weak"), and Z^0 (named by Weinberg), *requires* the existence of a Higgs field, with which the Higgs boson is associated, in order to give mass to these three bosons.

In chapter 9 we discuss the mysterious Higgs particle and the theory that predicted its existence. Chapter 10 describes the unification of the electromagnetic and weak forces, which relies on the existence of the Higgs. The search for the Higgs boson in the LHC is thus a major test of the Standard Model. If the Higgs is discovered, we will have found the source of the creation of mass in the universe, and we will also have provided a major piece of experimental evidence for the validity of the Standard Model. We will thus have completed this model.

Chapter 9

Hunting the Higgs

During much of the second half of the twentieth century, physicists were working hard to construct the Standard Model of particle physics: the quantum field blueprint of the universe. Their aim was to develop a kind of "periodic table" of the elementary particles that would show how all the leptons, quarks, and bosons fit together in a unified picture of nature and its forces, albeit without gravity. Remember that incorporating the modern theory of gravity, Einstein's general theory of relativity, within the Standard Model is still a distant hope. But as we know, small particles are not affected much by gravity, so the Standard Model has been successful in modeling and predicting particle interactions and phenomena.

While this model was still far from complete at midcentury because the weak force and its bosons were not yet understood, and neither was the strong force, a powerful idea had emerged about mass. According to this theory, right after the Big Bang there was no mass—only energy. Then, a fraction of a second later, something stunning happened: A symmetry of nature was suddenly broken, and the mechanism that broke that symmetry imparted mass to all the massive particles in the universe.[1] Thus, through work on the emerging Standard Model, the notion of the *creation* of mass in the universe was found to be inexorably linked with the concept of symmetry. The spontaneous breaking of a primordial symmetry that existed in the universe right after the Big

Bang thus created the mass of everything we see in nature today and enabled our own existence. This is certainly one of the most profound ideas in the history of science and philosophy. And what is even more shocking is that this notion can be experimentally tested by the LHC.

Let's put this incredibly powerful idea in perspective. Until the 1920s, there had been no proof that the universe had a beginning. Einstein believed that the universe was simply "there"—other galaxies were not known, and neither was the expansion of the universe. Then the discoveries through large telescopes of other galaxies, and of the expansion of space through the recession velocities of these galaxies, implied that the universe had a beginning: the Big Bang.

The Big Bang launched our universe—but how did the universe obtain its mass? To believe that all the immense amount of mass in the universe—billions of stars in each galaxy and billions of galaxies as far as our telescopes can see—was at one time all condensed at one point is much harder than it is to believe that *sheer energy*, rather than mass, was once packed inside a tiny amount of space that then expanded to form our universe.

Intuitively, this line of reasoning implies that mass was somehow created some time after the Big Bang—after the immense initial explosion. But until the emergence of quantum field theory, there was no mathematical way of "creating" mass out of energy. Quantum field theory, which wedded quantum mechanics with special relativity and led to the idea of matter and antimatter and their simultaneous creation out of sheer energy, actually uses mathematical *operators* that create and annihilate particles. The theory thus gave physicists the theoretical notion that mass can be created. But *how*?

The actual mechanism for creating mass after the Big Bang was the result of the work of Peter Higgs and, equally, a number of other physicists who worked in theoretical particle physics in the 1960s. The paramount idea behind the genesis of mass in the universe had to do with symmetry and how it broke shortly after the Big Bang.

We've seen various kinds of symmetries in earlier chapters. In addition to symmetries of nature that are intact, physicists can also perceive *broken* symmetries. There is a difference between a broken symmetry and a lack of symmetry. When there is no symmetry, nothing further can be said: There simply is no symmetry. But where a symmetry once existed, and somehow got broken, a physicist can clearly identify the remnants of the original symmetry.

Some symmetries are broken *spontaneously*—that is, in a natural way. For example, a crystal forms naturally. This is an example of a spontaneously broken symmetry, because space is perfectly symmetric in three dimensions before the crystal has formed, but once it forms,

Each of these crystals broke the symmetry of three-dimensional space when it chose a direction in which to grow

the direction of alignment of the atoms in the crystal breaks the original symmetry of space: By forming, the crystal defines a particular direction in space out of infinitely many prior possibilities.

But the best visual example of a spontaneous breaking of symmetry was given by the late Pakistani theoretical physicist Abdus Salam. Imagine that you are attending a dinner party. The guests are seated close together around a circular table. On your right, as well as on your left, there is a napkin. But because the guests are seated so close to one another, there is a perfect symmetry: napkin, person, napkin, person, and so on—all around the table. Not having read "Miss Manners," the guests don't know which napkin to pick up and use. So they sit for a while, in perfect symmetry. Then, spontaneously, one intrepid soul lifts up a napkin—on his right or on his left; let's say his left. At that moment, the *symmetry has been spontaneously broken.* Every guest now has no choice, but must follow suit and pick up the napkin on his or her left.

Note that with a broken symmetry, you can actually *see* that there once was a very well-defined symmetry, and that it got broken. A starfish with five limbs, one of which is broken, is still a starfish—you can see that there was once a symmetry here. It is not a perfect symmetry, of course, but that is true of all living things. Our faces are not perfectly symmetrical, and as many a tailor will gleefully point out to your spouse watching you try on a new suit, your arms are not of equal length. But compare a broken symmetry, such as that of a starfish whose limb has been broken, with the case of an asymmetrical piece of rock: The rock may have had no obvious prior symmetry. The theory behind the existence of the Higgs boson—the particle believed to endow particles with their masses—is all about symmetry and how it broke spontaneously in the very early universe, a fraction of a second after the Big Bang.

Appropriately—given that one of the main purposes of the LHC is to find the Higgs—three weeks after the dramatic first test of beams in the collider, the Nobel Committee in Stockholm announced that one-half of the 2008 Nobel Prize in Physics was to be awarded to the

Japanese American physicist Yoichiro Nambu of the University of Chicago (the other half was to be shared equally between the two Japanese physicists Toshihide Maskawa and Makoto Kobayashi, whose work on broken symmetries had predicted the existence of at least three generations of quarks). In the 1950s, Nambu inaugurated the theory of broken symmetries that eventually led to the hypothesis about the existence of the Higgs boson.

By 1959 Nambu had formulated a mathematical description of spontaneous symmetry breaking in particle physics. He derived his idea by analogy with what happens in superconductors: materials that under certain conditions, usually very low temperature, will conduct electricity with no resistance. (As we recall, the giant electromagnets that power the LHC are superconductors.)

Nambu started his work by studying an important 1957 paper written by John Bardeen, Leon Cooper, and Robert Schrieffer (BCS), which described how symmetries break spontaneously in superconductors. We know that as the temperature decreases in a superconductor, the natural mutual repulsion of the electrons is overcome by a weak attractive force, and at the point when the matter becomes superconducting, a symmetry of nature (in an abstract mathematical space) is spontaneously broken as electron pairs are formed. These are called Cooper pairs. Remember Wolfgang Ketterle's experiments with fermion gases, in which when the gas was cooled to a low enough temperature, the fermions paired up, with two atoms of opposite spins in each pair? Here we have the same idea.

Once electron pairs are formed, the electrical resistance of the matter goes to zero. Electric conductivity is now in a "super" state, hence the name "superconductivity." But something else happens too: An *energy gap* materializes. The energy gap is the threshold amount of energy that is required to excite an electron in a pair. When the temperature rises and the matter stops being superconducting, this energy goes to zero: The energy gap disappears.

It is as if the pair of electrons became a single entity. Each member of a pair has opposite spin from that of the other member; and their total spin is no longer fractional, because spin is additive. Since the pairs now have integer spins, they are now bosons, and under the right conditions can act as a single entity—a Bose-Einstein condensate. We can figuratively think of the pairing of the electrons as having "created mass"—a larger particle that can now congregate with others, thus forming something more "cohesive."

Regardless of how we visualize what is happening here, Nambu understood that the energy gap in superconductors *acts like mass* in the realm of particle physics. He derived the analogy using mathematics. When a piece of matter becomes superconducting, a natural symmetry is spontaneously broken by the formation of the Cooper pairs, and the energy gap materializes. In particle physics, by analogy, when the Higgs field undergoes a spontaneous symmetry breaking, mass is produced. The two processes are analogous.

Nambu derived this idea by looking at the equation describing the behavior of the Cooper pairs, and its solutions, and noticing that it bore a striking similarity to the Dirac equation of particle physics, and its own solutions. Once Nambu made a one-to-one correspondence between the mathematical terms describing the superconductor and those describing particle interactions, he found that the two equations were equivalent! The resulting theory entails the existence of a particle that today we call the Higgs. The name Higgs was attached to the hypothetical particle that imparts mass to the massive particles in the universe because of later work by the Scottish physicist Peter Higgs and several other physicists who independently published similar work at about the same time.

Imagine the universe permeated by a field called the Higgs field. When the universe was very young, the field had perfect symmetry. Then, just as happens in a superconductor, as the temperature in the universe dropped, the symmetry of the Higgs field was spontaneously broken, and now the field had a particular "direction" in the abstract mathematical space in which the original symmetry had existed.[2] The

Higgs field then produced "weight"—it gave masses to particles. And since a field can be excited, forming waves, and waves are also particles by the quantum principle of particle-wave duality, we got the Higgs particle!

The process can also be seen another way. When light (or electromagnetic radiation of larger wavelengths, also constituting photons) passes through a superconductor, it acts as if it has mass. This is a very curious effect: The light slows down; it acts "sluggish," as if it is composed of massive particles of matter. What Nambu suggested was that the same concept could be used in deriving mass for particles.

The idea is that all particles were born massless in the Big Bang, but space itself became something like a giant superconductor encompassing everything. Within this medium, some particles acquired mass, but the photon did not (ironically, the photon acts as if it were massive when it passes through an *actual* superconductor). So what is that superconductor-like medium that permeates all the space in the universe, giving particles their mass? It is what we call the Higgs field. The action of this field is carried out by a particle, a boson—since it is a force-carrying particle. We call it the Higgs boson.

But Nambu stopped his investigation here—something he deeply regrets because it could have led him directly to the explanation of the mysterious mechanism that gives all the massive particles in the universe their mass (except, perhaps, for the neutrinos where another mechanism may be at play). Nambu later said: "In hindsight I regret that I [hadn't] explored in more detail the general mechanism of mass generation for the gauge field. . . . I also should have paid more attention to . . . the present Higgs description."[3]

But a serious complication had arisen from an unexpected direction, and it stopped Nambu dead in his tracks. In 1962, the British physicist Jeffrey Goldstone, now at MIT, proved a theorem—with the help of Abdus Salam and Steven Weinberg—that showed that when a continuous symmetry is broken, massless bosons appear. So a theoretical physicist could indeed get the mass terms for the particles, as

Nambu had sought to do, by breaking a symmetry—but there would be a price to pay: the spontaneous creation of other particles, without mass. In fact, by Goldstone's theorem—which should be called the Goldstone-Weinberg-Salam theorem—we gain one massless boson for every degree of freedom lost from the original symmetry group. For example, if the original mathematical space is four-dimensional, and once the symmetry is broken the new mathematical space has, say, two dimensions, then *two* massless bosons must appear. This theorem caused huge problems, because no one had ever seen such massless bosons.

But how could physicists get rid of these nuisance bosons? No one knew the answer, and therefore Goldstone's theorem discouraged many experts from attempting to model the crucial concept of the creation of mass. And what is particle physics without mass for its particles?

The American Nobelist Philip W. Anderson derived essential elements of this theory while working in condensed matter physics. Anderson was born in Illinois in 1923; his father was a professor of plant pathology at the University of Illinois at Urbana-Champaign. Young Anderson loved spending his time hiking and canoeing and camping. Among his parents' friends were a number of physicists, who encouraged Philip to pursue the study of physics.

Anderson spent a year in Europe while his father was on sabbatical, and this experience gave him a deep appreciation for the wide world and for the exchange of ideas on a global scale. He excelled in high school and was accepted to Harvard University, where he obtained all his degrees, including a PhD in physics, which he got working under Julian Schwinger. He chose to work in condensed matter physics, in which the phenomenon of superconductivity is studied.

In 1957, the theory of Bardeen, Cooper, and Schrieffer, mentioned earlier, became important in the study of superconductivity. It explained this mysterious phenomenon by the behavior of electrons: It showed how the Cooper pairs form, act like bosons rather than fermions, and lead to the energy gap, which today we understand as analogous to mass in particle physics.

In field theory, excitations of fields are viewed as actual particles. Remember that in quantum mechanics a wave is a particle and a particle is a wave. So the excitation, the ripples in a field, constitutes a particle. Such particles, when acting within a crystal lattice or a superconductor, are called *phonons*.

Renormalization theory—the technique theoretical physicists use in order to remove the unwanted, unphysical infinite solutions that often plague their equations—was also used in condensed matter physics, just as it is used in particle physics. Condensed matter physics and particle physics thus became areas that use the same tools and the same ideas. "When the BCS [Bardeen, Cooper, and Schrieffer] theory came along," Anderson told me when I interviewed him about his work, "we finally understood field theory and were able to use it effectively in condensed matter physics."[4]

Anderson had been studying broken symmetries in the context of condensed matter when he came across the infamous Goldstone theorem. This result from field theory suggested that in condensed matter physics, as in particle physics, massless bosons should appear when symmetries were broken—that is, when normal matter spontaneously becomes a superconductor and the Cooper pairs form.

But Anderson saw no such massless bosons appear when the symmetry was broken in condensed matter, so he became convinced that the Goldstone theorem simply did not apply. He had thus realized something that other physicists could simply not see. In 1961, Anderson spent a year at Cambridge University in England. He talked to a number of physicists who were also working on broken symmetries but were discouraged by the fact that the Goldstone bosons kept blocking them from making progress in their theories. "There is nothing there, I told them," Anderson recalled. "There is no Goldstone boson!" One physicist responded: "If you feel this way, then you should write it up."[5] Anderson wrote his paper explaining why the Goldstone theorem fails and there are no massless Goldstone bosons and sent it to a journal in 1962.

It appeared in print in 1963. According to Anderson, Peter Higgs read this paper, and this led to his own article disproving the Goldstone theorem in 1964 (Higgs's references in that paper do not include Anderson's article; neither does his later paper, appearing in 1966). Anderson told me that "he [Peter Higgs] is an honest man and mentions my work; he tells it in an honest way."[6]

About the relationship of this work to Nambu's, Anderson said: "Nambu made the pion into a Goldstone boson; but he didn't understand how the electromagnetic field works in a superconductor. Here it is the BCS mechanism that is at work. There is no relativistic invariance."[7] The BCS theory is nonrelativistic (it doesn't use special relativity): in a superconductor, which is at a very low temperature, nothing moves very fast, so there is no need to talk about speeds approaching that of light and hence no need for relativity. And this is what allowed Anderson to show that the Goldstone-Weinberg-Salam theorem did not apply in condensed matter physics.

Shortly after Anderson made his great breakthrough showing that the theorem fails in superconductors, he was invited to lecture in the Soviet Union. This was in 1958, after a relative thaw in the Cold War, and physicists in the Soviet Union were eager to speak with him. A group led by the prominent Russian physicist Lev Landau was especially keen on meeting him. There was also another interested group, at the secret Russian accelerator at Dubna, but the Soviets withheld permission for Anderson to go there. Anderson arrived in Moscow in December 1958, during the infamous Russian winter with its short days. "We would have a long lunch," he recalled, "and after lunch, it was dark."[8]

At Landau's institute in Kharkov, Anderson spoke, and the standard hour allotted to a lecture passed. Landau was notorious for never allowing a speaker to go beyond an hour. But the Russian physicists were so interested in Anderson's work on symmetry breaking and the failure of the Goldstone-Weinberg-Salam theorem that Landau made an unusual exception letting Anderson speak well into the next hour.

Anderson was finally given permission to visit the Dubna group, but KGB agents seemed to be trailing him everywhere. At one point, inspecting the accelerator with a group of high-level dignitaries, he heard someone whisper "Psst . . . hurry!" He followed him quickly and in a deserted lecture hall met one of the physicists he had intended to see all along, one who had been interested in his work. After about twenty minutes of conversation, three men who "clearly were KGB agents" came over and led him away and back into the main group of visitors.[9]

Russian physicists understood the great importance of Anderson's work and valued his pioneering contributions. In the West, however, the spread of Anderson's ideas encountered much resistance because the fields of condensed matter physics and particle physics were not yet ready for new ideas. Furthermore, because relations between the Soviet Union and the West were not open, there was little exchange of information between scientists in the two political blocs; Anderson's work in condensed matter physics did not get the attention it deserved from particle physicists in the West. But independently, European and American physicists began to look at the problem posed to particle physics by the theorem of Goldstone, Weinberg, and Salam. And despite the fact that particles can move at very high speeds, and hence quantum field theory—which uses both quantum mechanics and special relativity—is used in particle physics, there is a way to avoid the disastrous consequences of the theorem.

Peter Higgs started to work on this problem in 1961, not long after he took a lectureship at the University of Edinburgh—where he has spent most of his long career and where he is now a professor emeritus. He had come there, to his native Scotland, after receiving a doctorate from King's College London, where he wrote his dissertation on the vibration spectra of molecules. Higgs had spent much of the previous six years traveling between Edinburgh and the theory group headed by Abdus Salam at Imperial College London and also visiting University College London.

When Higgs landed a permanent position at Edinburgh in October 1960 he didn't quite know which path his research should take. Then he read the stunning paper by Yoichiro Nambu, and it inspired him—as it did many other physicists around the world. He was now able to demonstrate mathematically that, using a gauge theory, no massless Goldstone bosons arise when a continuous symmetry is spontaneously broken.

A full month earlier, on June 26, 1964, the journal *Physical Review Letters* received a paper by the two Belgian physicists F. Englert and R. Brout of the Université Libre de Bruxelles proposing a similar solution to the same problem. And on October 12 of the same year, the same journal received yet another paper accomplishing the same task. This one was authored by G. S. Guralnik, C. R. Hagen, and T.W.B. Kibble, all of Imperial College London.[10] All these physicists demonstrated mathematically that the Goldstone-Weinberg-Salam theorem can be circumvented by using gauge theory ideas. Ultimately, it is the fact that gauge theories use local continuous symmetries (continuous symmetries that can be defined differently at different locations in space) that makes the Goldstone-Weinberg-Salam theorem fail to apply in particle physics, just as it does in condensed matter physics.[11] Therefore, the Standard Model can work with a boson that gives itself and other particles their masses once the symmetry of its field is broken when the universe cools a fraction of a second after the Big Bang.

Peter Higgs followed up his 1964 paper with another two years later, in which he explained the mechanism giving particles mass. Higgs has said: "The first paper merely says that there is no obstacle to this sort of theory. The obvious thing to do was to try it out on the simplest gauge theory of all, electrodynamics—to break its symmetry to see what really happens."[12] When Higgs completed his theoretical experiment breaking the gauge symmetry, he found what we now call the Higgs boson.

Physics Letters rejected Higgs's second paper. The editors felt that the paper had no real relevance to physics. So Higgs resubmitted it with one additional paragraph at the end, in which he hinted that the mechanism for producing mass described in the paper may have applications to the strong interaction—the action of the force holding the nucleus together (quarks were not known at that time). This got the paper accepted. Higgs has speculated: "This paragraph is perhaps why I get credited with the so-called Higgs boson."[13]

Higgs apparently feels that Anderson could have achieved what he and the others had, and has said: "Anderson should have done basically the two things that I did. He should have shown the flaw in the Goldstone theorem, and he should have produced a simple relativistic model to show how it happened. However, whenever I give a lecture on the so-called Higgs mechanism, I start off with Anderson, who really got it right, but nobody understood him."[14] A popular textbook on quantum field theory, by A. Zee, has a chapter titled "The Anderson-Higgs Mechanism."[15]

Understanding the Higgs and the Higgs mechanism has proved difficult for politicians, who are the ones to approve government spending on science projects such as building the LHC at CERN or the discontinued American project to build the Superconducting Super Collider (SSC) in Waxahachie, Texas. Part of the tunnel had already been dug near Waxahachie when Congress cut the $12 billion funding needed for this project in 1993. Now CERN's lower-total-energy LHC is the only advanced particle accelerator scientists can use to explore new physics.

In 1993, the United Kingdom's science minister, William Waldegrave, trying to understand what the Higgs was so he could support accelerator work, asked physicists to explain it to him on one side of a single piece of paper. He offered an expensive bottle of champagne as the prize to the physicist with the best explanation—meaning one he could understand.[16]

The physicists described the Higgs mechanism using many interesting analogies. The winning example was a variation on the following: Into a crowded cocktail party walks a famous star, immediately getting surrounded by people, who slow her motion through the crowd. This is how a particle (the star) gets "mass," while a less attractive person might be able to walk through the room without hindrance—hence no "mass" to drag her down. The less important person is like the photon, which does not acquire mass through the Higgs mechanism. The photon just zips through; a massive particle is hindered in its motion.

Now imagine that no new person comes into the room but rather that a rumor suddenly starts to spread. The congregating of heads as the rumor makes its way through the room is how the Higgs particle itself (through the Higgs field—the crowd in the room) acquires its own mass.

Michael Wick, a postdoctoral researcher from Munich, said he had a better analogy: "Think of a person jumping into a swimming pool," he said. "When his body hits the water, he feels his own mass."[17] John Ellis gave me another good analogy for the Higgs mechanism. "The Higgs field is like a snowfield," he said. "Imagine that you are walking through a snowfield. You certainly feel your 'mass'—you feel the drag of the snow on your feet. Now imagine that the snow melts away—you no longer feel your 'mass' and can walk easily."[18] The snowless field in this example is the perfectly symmetric universe very shortly after the Big Bang. Then, as the temperature cooled somewhat, "snow" appeared, spontaneously breaking the symmetry of the Higgs field; now the massive particles were given their masses.

Peter Higgs is hoping that "his" particle will be discovered in the LHC experiment. Stephen Hawking, on the other hand, reportedly does not believe that the Higgs boson will be found and has bet money against its discovery. Naturally enough, Hawking hopes for another major finding by the LHC: the discovery of a tiny black hole that will materialize for a split second and then evaporate according to the law of black hole radiation Hawking has theorized. For a massive black hole, a dead star, this takes an extremely long period of time, but for a tiny black hole it should take only a small fraction of a second. The two physicists are therefore on opposite sides in terms of their hopes and expectations of the LHC experiment at CERN.

Tension between the two began a few years ago. Hawking, a theoretical physicist known for his penchant for betting on what experimental physicists may or may not find, had placed a bet that the LEP, which preceded the LHC, would not find the Higgs particle. It appears that Peter Higgs was offended by this bet, a situation that was probably not eased when the LEP was shut down and Hawking won his bet.

At the beginning of September 2002, Higgs and other physicists were enjoying a celebratory dinner held at a restaurant off the Royal Mile in Edinburgh. The event was sponsored by Britain's Particle Physics and Astronomy Research Council to mark the opening night of a play based on the work of Paul Dirac. The British paper the *Independent* reported what supposedly took place at this event:

> . . . Stephen Hawking, the Lucasian Professor of mathematics at Cambridge University, was accused yesterday of receiving instant credibility because of his celebrity status. His accuser, a mild-mannered, retired scientist, is not the firebrand you might expect to find squaring up to a man who has been lionized as one of the world's greatest thinkers. Indeed, so modest is Professor Peter Higgs that he refuses in public to call the elementary particle named after him by its real name—the Higgs boson—preferring to give it the more impersonal moniker, the scalar boson.[19]

Scalar means that the field associated with the Higgs is represented throughout space as a single number (as opposed to a *vector,* which has magnitude and direction and is represented by a set of numbers). This property is related to the fact that the Higgs boson is a particle with spin equal to zero. For comparison, the photon, the boson of electromagnetism, has spin 1. So do the two W bosons and the Z boson of the weak interactions, which we will meet in chapter 10. And so does the gluon.

At the dinner, Peter Higgs supposedly said, "It is difficult to engage him [Hawking] in discussion, and so he has got away with pronouncements in a way that other people would not. His celebrity status gives him instant credibility that others do not have."[20] After the story about the dinner comments by Higgs was first reported in the newspaper the *Scotsman,* other newspapers in Britain, including the *Independent,* reported it, and Higgs promptly wrote Hawking to apologize. Hawking wrote him back saying he wasn't offended—but added that he still believed that the Higgs boson would not be found in later experiments, either—presumably at the LHC![21]

An unnamed physicist present at the restaurant where Higgs made his remarks has said, referring to Hawking and to Paul Dirac, whose work the dinner was celebrating: "Paul Dirac made a far bigger contribution to physics than Hawking, yet the public has never heard of him." And another added, "To criticize Hawking is a bit like criticizing Princess Diana—you just don't do it in public."[22]

Chapter 10

How the Higgs Sprang Alive Inside a Red Camaro (And Gave Birth to Three Bosons)

The real test of the idea of a Higgs particle, the real use of the powerful mechanism, the real theoretical proof that the Higgs boson has a *meaning* did not come from Peter Higgs or from any of the other physicists who developed ideas about how to circumvent the curse of the massless bosons. Ironically, it came from the same duo of brilliant physicists who had helped Goldstone prove his troublesome theorem in the first place: Steven Weinberg and Abdus Salam.

Both of the papers by Peter Higgs; the one by Englert and Brout; and the paper of Guralnik, Hagen, and Kibble were all concerned with showing that gauge theories constitute an exception to the Goldstone theorem. These physicists showed that within this setting a particular Lie group could undergo a spontaneous symmetry breaking and theoretically create mass without the attendant production of unphysical massless bosons. But none of these theoreticians had exhibited the actual Lie group whose symmetry is broken by nature—leading to the formation of massive particles. The actual mechanism that nature has chosen for creating mass was identified through the work of Steven Weinberg, Abdus Salam, and Sheldon Glashow. Their collective research not only brought us an understanding of the exact process that created the mass of the universe, it also presented another major accomplishment: the *unification* of two of the four forces of nature—the electromagnetic force and the weak force. Their achievement is one of the greatest triumphs of modern physics.

Teaching at Harvard, Julian Schwinger was very popular with graduate students and would often supervise as many as a dozen doctoral candidates at one time, leaving him little time to devote to each. However, he gave them such insightful advice that they ended up producing excellent research, and many of them have become leaders in their field. Nowhere was this truer than in the case of Sheldon L. Glashow.

In 1957, three years after the Yang and Mills paper on symmetry and gauge theories appeared, Schwinger published a seminal paper in the *Annals of Physics* in which he outlined his mathematical view of the interactions of particles and fields. Physicists began to believe that if gauge invariance (a continuous symmetry of nature whose properties are modeled by a Lie group) worked for electromagnetism and, as Yang and Mills had shown, could also be developed in the non-Abelian context of isospin, then perhaps the other forces of nature could also be successfully modeled using the gauge principle. By studying the theoretical behavior of both fermions and bosons, Schwinger was led to hypothesize the existence of a new particle responsible for mediating beta decay processes—that is, a particle that exerted the weak nuclear force, *causing* the neutron to emit an electron and an antineutrino (in the form of this process we've seen earlier).

But Schwinger's new particle could not easily be found, not only because the energy required was unattainable at that time, but also because his hypothetical particle would ultimately turn out to be no less than *three* particles (depending on the kind of weak-interaction process): one positively charged, one with a negative charge, and the third one neutral. What Schwinger also did in his article—something that may not have been noticed by his readers because it was done so subtly—was to suggest that electromagnetism and the weak interactions were perhaps two different manifestations of the action of what was once a *single* physical force.

But Schwinger made this prescient idea much clearer to his Harvard student Sheldon ("Shelly") Glashow. As Glashow recounted when I spoke with him in his Boston University office in May 2009, "Schwinger told me that he saw similarities between weak and electromagnetic interactions: Both were vectorial, and both were universal."[1] "Vectorial" means carried out by a vector boson—a boson with spin equal to 1. "Universal" means that they apply to several different kinds of systems. Given how busy Schwinger was, and how many students he had to supervise, this was enough. Glashow was on his way to investigating the similarities between two forces in physics that until that time had been viewed as different entities.

Sheldon Glashow was born in Manhattan in 1932. He had two brothers, who were eighteen and fourteen when he was born. His parents had immigrated to the United States in the early part of the century from Bobruisk, Russia, to escape the persecution of Jews by the Cossacks. His father worked hard and established a successful plumbing business in New York. Both parents believed in the value of a good education, and one of Shelly's brothers became a dentist, the other a physician. The youngest son, however, became interested in Galileo's laws of falling bodies and also performed some rather dangerous chemistry experiments in the basement of his family's house. Shelly Glashow went on to study at the famed Bronx High School of Science, which has launched many illustrious scientific careers, including that of Glashow's school friend and fellow Nobelist Steven Weinberg.

Both high school friends then continued their studies of physics at Cornell University, where they found other bright young science students and highly motivated professors in a stimulating environment full of ideas. After graduation, Glashow enrolled in the doctoral program at Harvard University in 1954, where he began his work with Schwinger.

Glashow wrote his dissertation on vector mesons in elementary particle decays, stressing Schwinger's idea that the weak force and electromagnetism should be worked into the same unified theory. In 1958, Sidney Bludman of the University of Pennsylvania suggested that a particular Lie group might be useful in modeling weak-interaction processes now that Yang and Mills had opened the way to such a theory.[2] Bludman was right, the weak force does follow a Lie group gauge structure, but he attempted no joint modeling of the weak force with electromagnetism. Once Glashow finished his dissertation in 1958, after the Bludman paper had been published, he and Schwinger began to work on a paper attempting to actually unify electromagnetism with the weak force. But, as he has put it: "Alas, one of us lost the first draft of the manuscript, and that was that."[3]

By then Glashow had won a postdoctoral fellowship from the National Science Foundation and planned to do his work at the Lebedev Physical Institute of the Russian Academy of Sciences in Moscow. He had strong support for this postdoctoral position from Russian scientists at the Institute and physicists at Harvard University. The Russians had a continuing interest in particle physics and were always eager to find out what their American counterparts were doing—especially since Communism had imposed strict restrictions on the flow of information in and out of the Soviet Union. Meetings between Russian and American scientists were seen as the only efficient way of exchanging information and reports on progress in the field.

Glashow decided to await his visa to the Soviet Union in Europe, so in September 1958 he sailed for England, from where he headed immediately to Copenhagen, hoping to work with some of the physicists from around the world Niels Bohr had gathered in his institute. "The Bohr Institute was remarkable," Glashow told me when we discussed his experience.[4]

Glashow recalled that in Copenhagen he often shared his lunch with Niels Bohr, who smoked his ubiquitous pipe; he also remembered the intellectual stimulation of the institute. During these months in Europe he also visited CERN, which had been founded only a few years earlier. "The big buildings you see at CERN today were already there," he said, "but I was given an office in the comfortable and spacious building that today only administrators use."[5] There was a strong connection between the new CERN and the older institution in Copenhagen: Both were established as international institutions to forge close cooperation among scientists from many nations. That was one of the strengths of such exciting places of research, where new ideas could be developed, extended, and explored. Both institutions provided coveted opportunities for a young physicist eager to launch a career.

While passing through Paris, Glashow met the already famous Murray Gell-Mann and came under his tutelage. The two men would continue to cooperate in work on physics problems for many years. In fact, after his return from Europe, Glashow would travel widely across America lecturing about Gell-Mann's work.

It was at the Bohr Institute in Copenhagen and at CERN over the period 1958–1960—while awaiting the Russian visa, which never came—that Shelly Glashow made his great contribution to the unification of electromagnetism and the weak force into a single gauge model. Glashow's paper, titled "Partial Symmetries of Weak Interactions," was accepted for publication in September 1960 and published in *Nuclear Physics* in 1961.[6] This work would lead to his Nobel Prize, shared with Weinberg and Salam.

While in Europe, Glashow enjoyed the good life: He skied, hiked, partied, and traveled throughout the continent "sometimes with a lady friend, sometimes alone," as he described it in his 1988 autobiography.[7] At one point, in Copenhagen, he was slipped a piece of paper with a name and a phone number by a young girl who lived in Sweden. So he called her, boarded a ferry to her town, and arrived there to discover she was a sixteen-year-old high school girl who had just broken up with her boyfriend and had asked Shelly to take her to a party to make the ex-boyfriend jealous. Glashow barely avoided a fistfight with the snubbed suitor and returned as quickly as possible to the more intellectually daring—but less dangerous—work he was doing in Copenhagen.

The Glashow model unifying the weak and electromagnetic interactions was a big step forward, but much more work had to be done to tie together these two forces. The mechanism by which the conjectured gauge bosons mediate weak interactions was not known, the nature of these bosons themselves was a mystery, and the process of unification was not well understood.

In the meantime, the works of Higgs; Englert and Brout; and Guralnik, Hagen, and Kibble all appeared during a single year: 1964. These results hinted at the existence of a mechanism, which we now call the Higgs mechanism, by which mass may be imparted to particles through the breaking of a symmetry of nature. But the thrust of these three papers was to show how symmetry can be broken without producing massless Goldstone bosons. Higgs's 1964 paper did not even deal with mass directly, mentioning the word only as part of the expression "massless bosons." The Englert and Brout paper and that of Guralnik, Hagen, and Kibble both mentioned massive gauge bosons, and so did Higgs's second paper, published in 1966. But they went no further.

Most important for the theory of how mass is obtained in nature, the gauge bosons that mediate the weak interactions were theoretically determined to be *massive*—not massless like the boson of electromagnetism, the photon. But neither Peter Higgs nor the other physicists who had reached the same conclusions about the symmetry breaking were able to take the necessary next big step forward and exhibit the actual mechanism that gives mass to these bosons. The insight to formulate the actual unified weak-interaction and electromagnetism gauge theory in the powerful context of the Higgs mechanism, imparting mass to the weak-force bosons, came from two of the greatest theoretical physicists of our time, Steven Weinberg and the late Abdus Salam.

Steven Weinberg was born in New York City in 1933 and, like Shelly Glashow and many other top physicists who had done so much important work during the twentieth century, was born to Jewish parents intent on providing the best education to their children and instilling in them the love of learning. Weinberg knew what he wanted to do in life from an early age. In his autobiography he recalled that his father had always encouraged his interest in science, and that "by the time I was 15 or 16 my interests had focused on theoretical physics."[8]

After graduating from the Bronx High School of Science in 1950, Weinberg went on to study physics at Cornell, graduating in 1954. That same year, he married a fellow Cornell undergraduate. Louise Weinberg is today a noted Texas attorney and law professor. Instead of continuing right away to graduate school, Steven decided to head to Copenhagen, following at a young age the path usually taken by elite older physicists, most of them already with doctorates and established research tracks. Weinberg wanted to get a head start in leading-edge science even though he had only a bachelor's degree. He applied for a Fulbright grant to study at Bohr's institute, but Bohr, advising the Fulbright committee that determined who got the grants, voted against him. "It wasn't that Bohr was against me personally," Weinberg explained to me in his office at the physics department of the University of Texas at Austin. "He simply did not think that foreign students who did not understand Danish would benefit from being at his institute. Though scientific meetings at the institute were conducted in English, classes were generally conducted in Danish."[9]

But Steven Weinberg was not one to be deterred by language problems. Not knowing at the time about Bohr's rejection, he also applied for a National Science Foundation (NSF) grant for the same purpose, and apparently Bohr was not consulted by the NSF. When the Weinbergs arrived in Copenhagen, Bohr welcomed them warmly, and Weinberg fondly remembers having a formal dinner with Bohr. "My wife even sat next to him, which was not good because Bohr mumbled, mostly in Danish, and she didn't understand most of what he said."[10]

In the cerebral environment of Bohr's international institute, Weinberg got to meet legendary physicists such as Werner Heisenberg and Paul Dirac; alas, they did not prove helpful. "Heisenberg only cared about his own work, and was not interested in what the young people were doing," Weinberg said with regret.[11] And Dirac, who had founded relativistic quantum field theory—the framework of modern particle physics—did not have the personality to help an aspiring young physicist.

But younger colleagues were very helpful to Weinberg, and he fondly remembers how David Frisch and Gunnar Källén helped him do real research in physics even before he entered a doctoral program. After a year in Copenhagen, the Weinbergs returned to the United States, and Steven continued on to a doctoral program at Princeton University, obtaining his PhD in 1957. It was a decade later, in the fall of 1967, while working as visiting faculty at MIT and performing the daily chores of a new father with a toddler daughter, that Steven Weinberg made his great breakthrough. It happened while he was driving from his home to his office.

Weinberg had been working for many months on an intriguing problem, trying to understand some of the properties of the strong nuclear force. He was looking at various Lie groups and how they act on the up and down quarks—this is called *flavor* symmetry, as contrasted with color symmetry, in the strong interactions inside the nucleus—and studying how this symmetry breaks down. It occurred to him that perhaps he could find a symmetry that was both local and global—meaning defined everywhere in space but also allowed to take different values at different locations. This property would make the symmetry very powerful. If he could do this, he knew, he would satisfy the theoretical requirements of a Yang-Mills gauge theory. Weinberg considered two possible Lie groups, but neither turned out to be correct, and he was getting nowhere.[12]

Then, while driving in his red Camaro to MIT one day, it occurred to him that this line of thinking could actually work in trying to understand *another* problem: that of the weak interactions. He understood that the methodology he had been developing could explain the behavior of the weak nuclear force, rather than that of the strong nuclear force. He realized that he had not been wrong: He had simply been working on the wrong problem!

In a moment of epiphany, Weinberg saw the appropriate Lie group and realized that this group could be used to *unify* the weak force with the electromagnetic force. (This advance was completely independent of the earlier work of Glashow.) Weinberg also realized that the symmetry he had just discovered was spontaneously broken, yielding a smaller Lie group, that of the circle, similar to the one used in electromagnetism.[13]

Since what he had was a Yang-Mills gauge theory, which enjoyed local symmetry, he saw that it magically gave rise to one massless boson—the usual photon of the electromagnetic theory, not one of Goldstone's infamous bosons—and *three* remaining massive bosons. These were the charged heavy particles W^+ and W^- and an additional neutral boson, which Weinberg named the Z^0. All three bosons *acquired their mass through the Higgs mechanism.*

As Barton Zwiebach has explained it to me, the Higgs field originally had four components.[14] When the symmetry was spontaneously broken, "the two Ws and the Z each ate one of them, gaining mass and leaving one Higgs for us to discover!"[15]

The electroweak theory was now complete, with all its parts falling perfectly into place. Weinberg was even able to estimate the masses of the massive vector gauge bosons. Experiments later showed that these are about 80 GeV for the two charged W particles and 91 GeV for the neutral Z.

I was fascinated by this process of looking at an extremely complicated physical situation, finding the exact symmetry that applies, and identifying the Lie group that would model it. I asked Steven Weinberg how he did it. "You just guess," he answered, "and then you see if your guess was right."[16] With the aid of his elaborately carved cane, he walked to the blackboard on the other side of his office and drew the Feynman diagram for beta decay. "I looked at the electron and the antineutrino coming out of the decay of the neutron," he explained, "and I thought that there must be some gauge theory that would link the two leptons," adding that his model should also work for quarks, but that he had concentrated on the electron and the antineutrino and hence had named his now-famous paper in *Physical Review Letters* "A Model of Leptons."[17] The Feynman diagram of the beta decay process is shown on page 172.

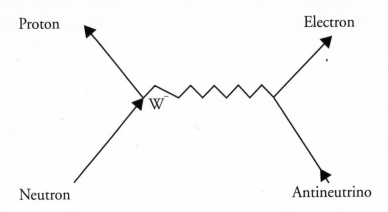

Proton   Electron

Neutron   Antineutrino

The beta decay process analyzed by Steven Weinberg (the arrow for the antineutrino is reversed, as an antiparticle is viewed as a particle traveling backward in time)

The paper is only two and a half pages long (with references and acknowledgments), but it is one of the most important physics papers ever written. Weinberg's paper was largely ignored for a number of years until physicists began to realize just how important it was. What Weinberg has achieved is the identification of the actual gauge symmetry broken spontaneously by nature, using the Higgs mechanism to impart mass to the three weak-force bosons while leaving the photon massless—thus creating mass not only for these three force carriers but also for all massive particles (perhaps except neutrinos), and breaking the weak force and electromagnetism apart as this primordial symmetry broke, destroying what was once a unified force very early in the life of the universe. And he was even able to estimate the masses of the three weak-force bosons. He did all of this in two and a half pages! "A Model of Leptons" is undoubtedly one of the most densely packed scientific papers in history. Weinberg's theory would be proved experimentally in most dramatic ways at CERN.

Abdus Salam was born in Jhang, in a farming district in what is now Pakistan but in 1926 was part of India. His father was an education official of the district. Abdus was an exceptionally good student and at age fourteen received the highest grades ever achieved by any student in the matriculation examinations held at the University of the Punjab. Based on his success, Salam was awarded a scholarship to study there and graduated with a master's degree in 1946. He then received a scholarship to study mathematics and physics at Cambridge University's Saint John's College, in England. His work there was so good that Cambridge University awarded him the Smith Prize for outstanding predoctoral contributions to physics.[18]

Salam then went on to take his PhD degree in theoretical physics from Cambridge. His dissertation on quantum electrodynamics was published in 1951. That same year, Salam returned to his native land, now called Pakistan after its tumultuous separation from India, and began to teach mathematics at Government College in Lahore. In 1952, he became the head of the mathematics department at the University of the Punjab. In 1954, however, he took a lecturing position at Cambridge.

Some years later, Salam went to Trieste, Italy, to found the Abdus Salam International Centre for Theoretical Physics (ICTP). The ICTP, which still exists after Salam's death in 1996, is an institution in which theoretical physicists from around the world gather to learn from one another while on leave from their home institutions.

Salam had a close friendship with Wolfgang Pauli. In 1956, after attending a conference in Seattle, Salam flew back to London on a U.S. Air Force transport plane full of servicemen and their crying children. He couldn't sleep on that long flight, so he derived in his head a theory on symmetry violations that involved the neutrino. Salam described what he did then:

> I wondered where I could go next and the obvious place was CERN in Geneva, with Pauli—the father of the neutrino—nearby in Zurich. At that time CERN lived in a wooden hut just outside Geneva Airport. Besides my friends, Prentki

and d'Espagnat, the hut contained a gas ring on which was cooked the staple diet of CERN—Entrecôte à la crème [steak in cream].[19] The hut also contained Professor Villars of MIT, who was visiting Pauli the same day in Zurich. I gave him my paper. He returned the next day with a message from the Oracle: "Give my regards to my friend Salam and tell him to think of something better."[20]

Salam was taken aback by Pauli's flat rejection of his idea about unifying forces of nature. But he consoled himself by reasoning that this reaction was a carryover from Pauli's distaste for Einstein's attempts to unify gravity with electromagnetism. Pauli had once said that these two forces "cannot be joined—for God hath rent them asunder."[21] But eight years later, after the Higgs mechanism had been proposed as a way of avoiding Goldstone's theorem, both Weinberg and Salam saw in it a way to unify electromagnetism with the weak force. In 1968, Salam wrote up his own version of electroweak unification in a paper, "Weak and Electromagnetic Interactions," in which he showed how to avoid the Goldstone theorem problem by spontaneous breaking of the local gauge symmetry, how mass of the gauge bosons is derived, and how the electroweak unification comes about.

Salam, in his Nobel talk, acknowledged the contributions of the many physicists whose collective works had brought us the "Higgs mechanism," which can be shown to give mass to the gauge bosons as well as other particles:

I shall not dwell on the now well-known contributions of Anderson, Higgs, Brout & Englert, Guralnik, Hagen and Kibble starting from 1963, which showed the way how spontaneous symmetry breaking using spin-zero fields could generate vector-meson masses, defeating Goldstone at the same time. This is the so-called Higgs mechanism. The final steps toward the electroweak theory were taken by Weinberg and myself (with Kibble at Imperial College tutoring me about the Higgs phenomena).[22]

Salam maintained his connections with Pakistan throughout his life. And he had two concurrent wives: one Pakistani, the other English. Salam married his Pakistani wife, Hafeeza, through an arranged marriage, and he had a son and three daughters with her. He then fell in love with a young woman he could not possess, and his despair over this hopeless situation was said to have pushed his decision to move to England. There he married his English wife, Dame Louise Johnson, an Oxford University professor, with whom he had a son and a daughter.

Apparently Salam kept his double marriage a secret for eight years, only once confiding to a friend, "You know, I have a second wife." Dame Louise gave an interview to a newspaper some years after her husband's death, in which she said: "By the time we met, he was already a distinguished scientist. I knew of his work and had been reading physics so I was very impressed by his ability. . . . There was something else in him too which brought us close. He grew up during the time of British rule and was therefore quite well-versed in English literature, especially the romantics. A little of that tradition did rub off on him."[23]

In 1979, both of Salam's wives came to Stockholm for the Nobel Prize award ceremony. Swedish officials were worried about an unprecedented problem of protocol: How could the King of Sweden receive two wives of one Nobel Laureate? Some even feared that the clash of cultural mores could create a diplomatic crisis between Pakistan and Sweden. Finally a solution was found: The two women were to be completely separated, never placed in proximity to each other, and seated at opposite sides of the reception hall.[24] The officials' tactful handling of this unusual situation preserved the peace and a diplomatic problem never materialized.[25] However, the audience at the award ceremony kept staring in the two opposite directions at the two women. Afterward, each woman held a separate celebration party for her husband.

Salam has been described as having had an irreverent and unorthodox approach to almost everything and "such a wonderful joie de vivre and his laughter, which most resembled a barking sea lion, would reverberate throughout the corridors of the Imperial College Theory Group."[26] Some have also described him as having had an intimidating personality; certainly his students would try to avoid a demanding professor ever hunting them down to assign them tasks.[27]

But how do we know that the Weinberg-Salam theory actually works? The proof that the Weinberg-Salam model is renormalizable— meaning that it can be freed of the problem of producing nonsensical infinite solutions, something that has plagued many models in physics— was provided by Gerard 't Hooft.

Gerard 't Hooft was born in Den Helder, Holland, to a family that belonged to the Netherlands' scientific elite. His granduncle, Frits Zernike, had won a Nobel Prize for scientific work that enabled him to invent a new kind of microscope. He then took his invention and demonstrated to biologists how they could use it to see images of living cells.

Zernike's sister was 't Hooft's grandmother; she married another well-known scientist, her former professor Pieter Nicolaas van Kampen. Her son, Gerard 't Hooft's uncle, was Nicolaas Godfried van Kampen, a professor of theoretical physics at Utrecht, who became his nephew's mentor. Already as a schoolboy, when Gerard was asked what he wanted to be when he grew up, he answered: "A man who knows everything." He did not know the word "professor" yet, but he already knew what it meant to be a great one.[28]

The physics teacher at 't Hooft's high school used a textbook that he and a colleague had authored. Gerard 't Hooft found an error in the physics and, with the help of his famous uncle, Professor van Kampen, showed the teacher he was wrong. As a result, "this page does not appear in later editions of the book."[29]

Under the tutelage of his uncle, 't Hooft studied theoretical physics at the University of Utrecht. There he came under the influence of a young professor of particle physics, Martinus ("Tini") Veltman. Tini suspected that 't Hooft's stellar grades at school were due to his family's prestige and wanted to test the abilities of the new undergraduate he was advising.[30] So Tini gave Gerard the paper by Yang and Mills on non-Abelian gauge theories. This was clearly no undergraduate stuff since the mathematics, the physical ideas, the equations, and the derivations can challenge even the best-prepared doctoral student. But the beginning student from an elite scientific family surprised his professor by devouring the paper and understanding it completely. The two men began to discuss Yang and Mills seriously.

Gerard 't Hooft found the Yang and Mills article to be "a brilliant paper." It was beautiful, meaningful, and unique. But it was also useless. "It describes particles which do not exist in nature," Veltman told his student, "but in some modified form, they might."[31] What modified form? Gerard asked himself. There was, he recognized, "a lot of confusion concerning the so-called Goldstone theorem. This was an example of people depending too much on abstract mathematics and not reading the small print," 't Hooft explained.[32] He never believed in the existence of massless bosons coming out of the Goldstone theorem, he told me when I interviewed him in May 2009.[33]

In 1969, Gerard 't Hooft became Veltman's doctoral student, and Tini gave him a choice of problems to work on for his doctorate. But none captured his imagination more than the very difficult problem of the renormalization of Yang-Mills models. In the summer of 1970, 't Hooft went to the summer school on theoretical physics held at Cargèse, in Corsica—a location chosen for the conference because it gets on average more sun than any place in France. Some of the best theoretical physicists were meeting there, and this was the place where 't Hooft first made progress on the problem of renormalizing the Yang-Mills gauge theories so that the infinities could finally go away for good and the model would make perfect physical sense.[34]

Veltman and 't Hooft then worked together on making the gauge theories work. Veltman believed in doing much of the calculations using a computer, which was a novel approach in theoretical physics at that time. He worked with the massive early computers that were housed in computer centers and were fed decks of cards as their form of input (this was years before keyboards that enter data directly into a computer were invented).

Eventually Veltman and 't Hooft were able to show that the Yang-Mills gauge theories were renormalizable. This was a huge breakthrough in theoretical particle physics, and they were jointly awarded the Nobel Prize in 1999. In the words of another Nobel Prize winner, Abdus Salam, quoting the late American particle physicist Sidney Coleman: "'t Hooft's work turned the Weinberg-Salam frog into an enchanted prince."[35] But the theory required strong experimental proof in order to be fully established. This was soon to arrive through a stunning discovery at CERN.

In 1973, André Lagarrigue discovered the first piece of evidence in favor of Weinberg's theory. CERN had commissioned the French

The famous bubble chamber called Gargamelle, on display at CERN, with physicist Paolo Petagna in the background; behind him is the Big European Bubble Chamber

nuclear research center at Saclay, located southwest of Paris, to build it a powerful bubble chamber that could be used in various experiments conducted through the CERN accelerators. French engineers then built a large bubble chamber about the size of a small research submarine and looking just like one, with a steel body, now painted orange, and many portholes. They named it Gargamelle, after the mother of the gluttonous giant, Gargantua, in the satire by the sixteenth-century French writer François Rabelais, *Gargantua and Pantagruel.* This bubble chamber is indeed gargantuan: It weighs 25 tons, and during the time it was operational, it was filled with 18 tons of either Freon or propane.

CERN had a list of experiments to be conducted through the use of Gargamelle, starting in the early 1970s. In a meeting in May 1972, a committee of scientists managed to convince the directorship of the laboratory to allow them to carry out an experiment that had a low priority, only eighth on the list of possible uses for the bubble chamber: looking for the doubtful "neutral current." A current, as we know from everyday life, is the flow of electrically charged particles through a wire or some other medium. It is an interaction governed by the electromagnetic force. A neutral current would be the flow of neutral particles, meaning ones without an electric charge, created by the action of the weak nuclear force. At that time, few people believed that such currents existed. But Weinberg's neutral boson, the Z^0, would be the mediator of such neutral currents, if indeed they existed in nature.

André Lagarrigue and his collaborators at CERN set to work right away and created very powerful beams of neutrinos and antineutrinos, which emanated alternately from an accelerator and were focused right into the Gargamelle chamber. As we know, neutrinos are very reluctant to interact with matter, but the flow of these particles from the accelerator was so high that every once in a while the interaction of a neutrino or antineutrino with the liquid inside Gargamelle was recorded through the trail of bubbles it created. Once in a while, the neutrino itself survived. When this happened, it constituted proof of a neutral current and hence gave indirect proof of the existence of Weinberg's Z boson.

But where were these bosons? Somebody had to actually find them. The neutral boson Weinberg predicted, the Z^0, and the two charged bosons, the W^+ and the W^-, were discovered through proton-antiproton collisions inside CERN's then-new SPS accelerator by a team led by the Italian physicist Carlo Rubbia, with important work done by the Dutch physicist Simon van der Meer. The W bosons carry a charge of +1 or −1, depending on the kinds of interactions they are involved in (the beta decay process that produces an electron and an antineutrino is mediated by the W^- boson, and a "reverse" type of beta decay process, in which a positron and a neutrino are created, is mediated by the W^+ boson). The W^+ and the W^- are each other's antiparticles. Both W bosons are very heavy—which is why they were hard to detect—with mass of about 80 GeV each. The third boson of the weak interactions, Z^0, which has zero electric charge (and is its own antiparticle), and thus mediates reactions with no electric charge, has an even higher mass: 91 GeV (Weinberg had estimated it at more than 80 GeV). This is a hundred times the mass of a single proton.

Discovering the W and Z particles was therefore a great challenge. The successful outcome was the result of considerable effort at CERN by 135 scientists working together day and night in two groups, one of them led by Rubbia. The scientists analyzed a huge amount of data. The charged W bosons (each of which lives for a mere 3×10^{-25} seconds before disintegrating) were found in January 1983, and the neutral boson, Z^0 (which also lives for only 3×10^{-25} seconds before it decays into other particles) was discovered four months later, in May 1983. The Nobel Committee wasted no time and in an exceptional move awarded Rubbia and van der Meer the prize the following year.

The present LHC work requires much more intensive data analysis than the quite appreciable computational effort that was needed to prove the existence of the bosons of the weak interactions, but computer technology has made huge strides in the last two decades; even so, what is required for the LHC work involves thousands of computers connected on a grid through the World Wide Web to perform distributed computing at locations around the world.

Out of the first billion collisions carried out by Rubbia's team and a second group of scientists working with the SPS, a total of six W particles were discovered. Later work analyzing another billion collisions revealed five additional W particles.[36] By now, tens of thousands of W bosons and millions of Z bosons have been discovered through analyses of particle collisions in large accelerators. The discoveries at CERN of the three bosons that mediate the action of the weak force constituted a powerful proof of the electroweak unification theory of Weinberg, Salam, and Glashow. It also solidified physicists' universal trust in the Standard Model, of which the electroweak sector is a part. Physicists now became convinced of the paramount importance of finding the Higgs—the mysterious particle responsible for imparting mass to the three new weak-interaction bosons and to other massive particles.

Even though it does not incorporate the effects of gravity, the Standard Model of particle physics is universally considered one of the greatest successes of modern science—many physicists consider it *the* greatest model in physics—because it accurately predicts many new phenomena and explains a host of known ones. The Standard Model shows how the particles relate to each other in each group and across groups; and how the forces of nature, the electromagnetic, the weak, and the strong, act on the particles through the mediation of special-purpose particles, the force-carrying bosons. But, as we know, the Standard Model does not explain everything. In addition to not addressing the effects of gravity, it also does not explain the particle masses, and it does not allow for perfect unification of the three forces during the early life of the universe.

When the Standard Model is extrapolated back in time to very shortly after the Big Bang, the three forces of the Standard Model do not match at a single point—and physicists think that they should unify in that realm of the extremely high energies that followed the Big

Bang. Finally, the Standard Model also does not explain the existence of dark matter and dark energy—the greatest mysteries from space.

The Standard Model is, however, the best working model of particle physics we presently have, and the achievement of Glashow, Weinberg, and Salam, the *unification* of two of the four forces of nature, is an immense advance in modern science. The two forces, electromagnetism and the weak force, seem on the surface to be very different: one is a force acting on electrons and mediated by a massless boson, the particle of light we call the photon. It is a force that acts through distance: Remember the magnet that can affect an object from afar, or think of lightning that can strike a point on the Earth from high up in the atmosphere. The other force, the weak force, acts at very short distances, inside the nucleus of an atom. And this force is mediated by very massive bosons: the two W's and the Z. And yet, deep theoretical similarities between the two forces prompted these three Nobel physicists to believe, and ultimately to prove, that electromagnetism and the weak force are really two aspects of one unified force, whose symmetry broke in the early universe.

The electroweak unification, as it is called, gave new hope to physicists that perhaps the final theory of physics would successfully unify *all four* forces of nature. Furthermore, extrapolations of the strengths of the four forces, carried back in time toward the Big Bang, indicated that the forces "want" to come toward a single point at the energy levels right after the Big Bang—but they can't quite make it. This has prompted some physicists to hypothesize that perhaps the Standard Model is not the final word in physics and that another theory should be sought: one that would match the forces better and hence unify them during the epoch right after the Big Bang. We will discuss the two theories that hold the most promise of achieving this goal in chapters 11 and 12. Chapter 11 is the first to take us beyond the Standard Model, and there we will explore phenomena the LHC may discover that could shed light on and elaborate on some of these wild, untested models at the forefront of science.

Chapter 11

Dark Matter, Dark Energy, and the Fate of the Universe

Einstein's general theory of relativity is universally considered the model of a perfect theory in physics: Einstein's equations have a quality that physicists call beauty, or elegance. What does this mean?

Einstein's theory describes a very complicated space-time reality in stunningly concise equations with a minimum of parameters. The theory thus agrees with the principle known as Occam's razor: The simplest theory is likely to be the right one. Or, as Einstein himself famously put it, a theory should be "as simple as possible, but not simpler."

There is also an overarching, powerful symmetry of space—called general covariance—which assumes that the laws of physics do not depend on whatever coordinate system we use to label space and time. For example, the distance from some ideal city's Thirty-fourth Street and Second Avenue to Nineteenth Street and Fourth Avenue should not depend on the labels we use for the streets and avenues (assuming they crisscross at right angles and are equally spaced). This symmetry underlies the logic of Einstein's model, and it lends it elegance; it is also necessary for explaining the laws of physics. Following Einstein, the idea of elegance has taken on a life of its own in physics, not only because it produces equations that are aesthetically pleasing, but also because elegance seems to work in explaining nature. A pleasing equation tends to be correct: Nature loves beauty.

The success of the next great model in physics, the Standard Model—also a theory based on the idea of symmetry—has energized physicists to look for more models that are mathematically and aesthetically pleasing, with the underlying belief that such models are likely to be found correct in describing nature. But does nature *always* follow these principles? In this chapter and the next we meet several relatively new physical theories—developed mostly during the last third of the twentieth century (although their roots go back to the earlier parts of the century)—which are considered beautiful, exciting, and fun to contemplate and pursue. But none of these theories has yet found a shred of supporting evidence from experiments.

On the other hand, there are also *physical phenomena* that so far have eluded all our attempts at explanation. Therefore there is a strong need to try to match the elegant models with the unexplained phenomena, and physicists hope that at least some of these mysteries might be solved by one or more of our beautiful new theories. The ultimate test of at least some of these models may lie in the work of the LHC.

We have understood since the 1920s that the universe is expanding. But not much has been known about how the speed of this expansion is changing through time. Many physicists had assumed that because the force of gravity extends indefinitely in space, the mutual attraction of all the mass in the universe should eventually slow down the expansion caused by the Big Bang, making the universe recollapse on itself sometime in the very distant future. Then perhaps another Big Bang would occur, in what could be part of a perennial cycle of birth, death, and rebirth.

Then in 1998 this hopeful philosophy of nature and its ability to regenerate itself was shattered to bits. Two teams of astronomers, one at Berkeley, headed by Saul Perlmutter, the other at Harvard, headed by Robert Kirschner, independently announced their surprising findings from studies of the rates of recession of distant galaxies. They reported that, rather than slowing down, the expansion of the universe is actually *accelerating*. This means that—unless a miraculous force somehow intervenes to override the acceleration—the universe will expand forever. Eventually, in the very distant future, it will become extremely diffuse and die once all the stars run out of their nuclear fuels.

These findings constituted big news in science—no one had expected them. The mathematical catalyst of the accelerating expansion was identified as a long-discarded constant, called the cosmological constant, or lambda, which Einstein had originally used in his model of the universe in 1917. He did so because, while his modeled universe wanted to expand, astronomers at that time told him that the universe was stationary. So Einstein added to his equation a term he called lambda (after the Greek letter he used to denote it) to keep the universe from expanding—thus missing the opportunity to actually *predict* theoretically the expansion of the universe!

When Einstein was made aware of the 1929 discovery by Vesto Slipher, Edwin Hubble, and Milton Humason of the expansion of space, he threw away lambda with disgust and famously exclaimed: "Then away with the cosmological constant!" Ironically, to model an *accelerating* universe, one could use a term such as lambda in Einstein's equation, and instead of keeping the universe from expanding, it would actually accelerate it.

Whether or not it is modeled by lambda, scientists could not identify the force responsible for accelerating the expansion of the universe. The inescapable conclusion has been that there is an unknown, mysterious form of *energy* that permeates all space—there is *something* out there, all over space, a kind of shadowy power that constantly "pushes" on the very fabric of space-time, overcoming the force of gravity and pressing space outward. It is a bizarre kind of force that can't be seen or felt other than through its action of accelerating the expansion of the universe. This unseen energy is called *dark energy*.

The discovery of dark energy brought back to the fore another mystery about the universe, one that has puzzled scientists ever since its discovery in 1933 by the Swiss American astronomer Fritz Zwicky of the California Institute of Technology. After carefully studying the collective gravitational pulls of galaxies on their own stars and on other galaxies in their respective clusters, Zwicky reached the inescapable conclusion that it just didn't add up: There was no way that the mass he estimated to exist in the galaxies based on all the stars and dust in them was anywhere close to being enough to hold them from flying apart! He concluded that the universe must contain much more mass than can be seen through our telescopes. This mysterious missing mass is now called *dark matter.*

All the later astronomical studies over the decades, using telescopes of ever-increasing power and methods of analysis that enjoyed equally impressive gains, have not been able to resolve the persistent discrepancy Zwicky had discovered between the total perceived mass and the total mass inferred by the collective gravitational pull of galaxies. So where is all the missing mass of the universe? Dark matter has so far *only* been detected through its strong gravitational effect on ordinary matter. It does not seem to interact electromagnetically, as normal matter does. We don't see it and we don't feel it—but we know that it's out there because it is necessary for balancing all the gravitational forces needed to hold together the stars and the galaxies.

Scientists now believe that a full 96 percent of the mass-energy (remember that by Einstein's formula the two are equivalent) in the universe is unseen; 73 percent is dark energy; 23 percent is dark matter; and almost all of the *remaining* 4 percent of the mass-energy is gas and dust. Only 0.4 percent of all the mass-energy in the observable universe is in the form of light-emitting stars.[1] Dark matter thus comprises a full 85 percent of all the *matter* in the universe.

Calculations show that the missing matter is unlikely to be hiding in black holes or to be accounted for by neutrinos, which are very abundant and don't interact much with other matter. Their total mass does not add up to anywhere near the missing amount. At present, nobody knows anything about the nature of the immense amount of missing mass in our universe.

The twin mysteries of the energy-mass of the universe—dark matter and dark energy—are now viewed with increasing urgency by scientists because these mysteries have defied all attempts to uncover their nature. So the search is on full-swing to find candidates for the dark matter that appears to make up close to a quarter of everything in our universe, and to try to explain the dark energy. The need to solve these puzzles has brought in particle physicists—experts in the world of the very small—for help to the astronomers, astrophysicists, and cosmologists who have been stymied in their attempts to find solutions to these two major problems in the realm of the very large. Particle physicists responded to this challenge by trying to explore physics beyond the Standard Model.

In their quest for dark matter candidates, physicists have come up with many possibilities, including *exotic matter:* bizarre, imagined particles, rather than entities based on anything seen in the physical world around us. Some potential candidates for dark matter that physicists have thought up are so mysterious that they are denoted by letters such as "D" for "dark" or "X" for "unknown" and by names such as Q-balls (for blobs of matter), axions, and saxions. For the moment, these exotic entities live only in the imagination; but if any such bizarre matter is identified through discoveries in the LHC, it would certainly change our view of the structure of the universe.

Some scientists have suggested that dark matter could be seen through its interactions: A dark matter particle may annihilate when it meets a dark matter *antiparticle,* releasing energy that we can detect. As mentioned earlier, the PAMELA (Payload for Antimatter Matter Exploration and Light-nuclei Astrophysics) satellite, launched in 2006 by a European consortium of scientists, looks for antimatter from space. It also searches for evidence of exactly this kind of interaction—the annihilation of dark matter particles as they meet dark matter antiparticles.

According to the leading particle physicist Gordon Kane of the University of Michigan, recent PAMELA data are consistent with such annihilation processes and may provide a secondary piece of evidence in support of the existence of dark matter.[2]

The greatest hope for results from the LHC, other than the discovery of the Higgs, is that the collider may help identify some of the missing mass of the universe through a discovery of one or more of a new set of particles predicted to exist by a particular kind of theory that extends the symmetry of the Standard Model.

Symmetries within the quark group brought us quantum chromodynamics and the prediction of new particles; symmetries of leptons brought us the Weinberg-Salam-Glashow unification of the electroweak force; and the joining together of three kinds of symmetry brought us the full Standard Model. So, with such great successes for the idea of symmetry in nature, a number of physicists have asked themselves, *Couldn't the same idea be extended further?* In particular, could the Standard Model symmetry be enlarged to directly tie together more of the particles and thus predict the existence of "new" ones?

Let's look at the groupings below:

Quarks   Leptons
               Bosons

There are two directions that physicists have considered for extending the Standard Model. One approach was to try to find a symmetry that would bring together *both* quarks and leptons; another was to look for a symmetry that would link all the *fermions*—quarks and leptons—with the *bosons*.

Abdus Salam and his colleague Jogesh Pati wrote an article in 1974 suggesting that the color charge of quarks could be extended to *four* colors instead of the usual three, and that the fourth color would allow the incorporation of leptons in the quark model. About the same time, Sheldon Glashow and Howard Georgi at Harvard started to search for a Yang-Mills gauge group that would allow them to put together mathematically all the quarks and leptons in one model. The pair settled on a large new Lie group, which *contained* the Standard Model inside it, and also linked the quarks and leptons in a single symmetry.[3] Physicists call such an approach a *grand unified theory.* But don't let the word "grand" mislead you—the theory is only grand in a limited sense: It unifies quarks with leptons. The Georgi-Glashow model was the first grand unified theory.

A major implication of grand unified theories is that quarks are transformable into leptons. In particular, a theory such as this one implies that the proton will eventually *decay* into a lepton (one with a positive charge, by electric charge conservation, so an antiparticle). This interaction has a grave implication about the fate of the universe: Given enough time, *all the hadronic matter in the universe would disappear!* If this theory is right, then in the very distant future, protons would break down, causing nuclei to disintegrate, and eventually there would be no atoms in the universe—just lonely electrons, positrons, and neutrinos flying in the ever-expanding emptiness of space.

But proton decay has never been detected experimentally, and scientists have been looking for it for many years now by observing large tanks of pure water placed deep underground, in search of radiation that could not be explained by anything else. The Super-Kamiokande project in Japan looks for proton decay, in addition to neutrino oscillations. Scientists working on Super-Kamiokande have recently reported that if proton decay is real, it has to take place after a time interval of at least 8.2×10^{33} years (the previous estimate was at least 10^{32} years). So it looks like we are safe from disintegrating anytime soon.

At present, we don't know if the proton can decay, since evidence for such a process is only theoretical, and depends on a mathematical model with special, untested assumptions. The speculation that the seemingly very stable proton may decay after a very long time has been around for many years. Maurice Goldhaber, who directed the Brookhaven laboratory and had hopes of making an experimental discovery about proton decay said as early as a quarter century ago: "May Proton live forever, but if it dies, let it die in my hands!"[4]

The second kind of extension of the Standard Model explored by physicists has been an effort to encompass both the bosons and the fermions in one symmetry group. This idea is called *supersymmetry*—it is a symmetry that is larger than that of the Standard Model and that includes the Standard Model as a subgroup. Supersymmetry is often abbreviated as SUSY.

The idea of supersymmetry was developed in the Soviet Union in the 1970s, but the work of the Russians did not become known in the West for many years. In 1971, Yuri Gol'fand and Evgeny Likhtman began to construct a supersymmetric field theory, and two years later another pair of Russian physicists, Dmitri Volkov and Vladimir Akulov, theoretically found the first broken supersymmetry. Unbeknownst to the Russians and vice versa, the Italian physicist Bruno Zumino, now an emeritus professor at the University of California at Berkeley, and his late Austrian colleague, Julius Wess, developed the idea of supersymmetry at about the same time that the Russian physicists were working on this problem.

Julius Wess had studied at the University of Vienna, where he came under the influence of Erwin Schrödinger. Bruno Zumino was already a professor when he met Wess, at the time still a graduate student. Together they started to work on symmetries in physics, devising what many physicists today consider an especially elegant model of nature. Some call it "beautiful SUSY."[5]

In one of its early theoretical successes, SUSY solved the problem of the varying magnitudes of the four forces of nature. As mentioned earlier, under the Standard Model, the three forces don't quite match in the distant past, a time right after the Big Bang, but they do under supersymmetry. And gravity, when extrapolated back in time, may also match with the other three forces.

Because the three (or the four, under certain assumptions) forces of nature *match* under supersymmetry in the early universe, supersymmetry offers us a unification of the forces. The forces are seen to go their separate ways only after the universe has cooled down and original symmetries have broken spontaneously; but if we use the SUSY model, we can see that a great symmetry was once there. The early unification, followed by symmetry breaking, lends the model much elegance as a theory that ties together all the forces around the time of their creation.

Another great advantage of supersymmetry is that this theory is well behaved theoretically. It requires less work to renormalize it—infinities are easier to remove from solutions than they are under the Standard Model.

Two weeks before he died in 2007, Julius Wess spoke at the international conference on supersymmetry, SUSY07, held that year at the University of Karlsruhe in Germany. He was very hopeful about the start date for the LHC, which at that time was planned for 2008. Unfortunately, Wess did not live long enough to see whether the LHC would verify his and Zumino's theory. In his talk at the conference, Wess stressed the idea of symmetry and how it led him and Zumino to supersymmetry:

In some way or another, symmetry is perceived by everybody. I think it is worth mentioning that about thirty years ago, there was strong interest in experimenting with apes to see how much they were able to learn. One objective was to see how apes would learn to paint. In one of these experiments one dot was made on one side of a piece of paper and the ape would then try to make a dot on the other side to balance it symmetrically. That's exactly what we are doing in physics.[6]

The SUSY idea is that each boson has a fermion partner—another dot symmetrically placed on the other side of the paper, as it were. And equally, a fermion has a boson partner. These are called *superpartners*. But since there are different numbers of fermions and bosons, more particles must be "invented" and added to the model in order to match up the two groups completely. Some of the as yet undetected particles needed for the model are excellent candidates for dark matter.

The gauge bosons of the Standard Model have fermion partners in the supersymmetric world. They are called gauginos. For example, the gluon has a superpartner called *gluino*; the W boson's superpartner is the *wino* (pronounced "WEEno"). The *chargino* is a charged particle that is a quantum mixture of particles; and the *neutralino* is a neutral particle that is also a quantum mixture. As we've seen when we discussed the neutrino, the suffix "ino" is the Italian diminutive (a cell phone in Italy is called a *telefonino*) and it seems that this is the source of the naming scheme—perhaps an extension of Fermi's idea behind naming the neutrino. And, wouldn't you have guessed it, the superpartner of the Higgs is the *higgsino*.

Other superpartners are named by adding a prefix "s": There are sleptons and squarks—they are the superpartners of the leptons and the quarks of the Standard Model. In particular, the bottom quark's superpartner is the sbottom; the top quark's superpartner is the stop; and in the slepton sector, we have, for example, the stau. According to one theory, the squarks and the gluinos are perhaps the heaviest superpartner particles (sparticles).[7]

Within the SUSY family of models, some are closer to the Standard Model, while others are more elaborate and hence more distant from it. The supersymmetric model that most closely resembles the Standard Model of particle physics is called the minimal supersymmetric Standard Model, or MSSM. This powerful theoretical construct in modern physics has many adherents and practitioners. Its limitation is that it is not likely to explain much of the missing matter of the universe. To explain the source of dark matter, more extreme assumptions are required. A further model is the NMSSM—the "next to minimal supersymmetric Standard Model." Both of these somewhat minimalist extensions of the Standard Model do include new particle candidates that might be confirmed in LHC collisions. The more complicated supersymmetric models, however, require higher beam energies for observing the phenomena they predict, and these theories are therefore less likely to be validated. At yet higher energies still, we find the string models, discussed in chapter 12.

The dark matter problem is so pressing for physicists that as the LHC begins its run, there is as much concern with discovering the source of the dark matter as there is about trying to discover a Higgs particle. Since the supersymmetric partners—even when found—may not add enough to the collective mass of all the known particles of the Standard Model to account for the full extent of what's missing, some scientists have come up with even more outlandish theories.

Presentations by physicists working at the leading edge of particle physics, given at a number of high-level professional congresses held on the eve of the LHC opening day, outlined the ways in which supersymmetry could find its confirmation through the work of the giant collider. I describe below some details of theories presented at these conferences, just to give you an idea of the flavor of the most advanced research in theoretical particle physics and the richness of the work of leading researchers.

The *graviton* is the hypothesized boson that mediates the action of the force of gravity. It has never been observed. If the graviton exists, then according to theory it must have a spin of 2. Recall that the W and Z bosons and the photon and gluon all have spin 1; and the Higgs, if it exists, has a spin of zero.

There is a theory called *supergravity*, which is an extension of supersymmetry that includes gravity. This theory predicts the existence of the *gravitino*: an analog of the superpartners of the W and Z bosons of the Standard Model, which is the superpartner of the graviton. According to a recent assessment, the gravitino may be a good dark matter candidate.[8] "The gravitino may be a natural dark matter candidate. But we have no idea about what its mass may be," said Wilfried Buchmüller of Hamburg, in a presentation at the supersymmetry conference SUSY09 in Boston.[9]

Another excellent candidate for dark matter particles within the supersymmetry theory is believed by some to be the *gluino*. Recall that this particle is the analog in the supersymmetric world of the gluon of our world. Physicists believe that gluinos will be detected by decaying into jets and staus.[10]

According to Raman Sundrum of Johns Hopkins University, nature has a "Dark Sector," with its "Dark Particles," which are separate from the supersymmetric partners all lumped in the "SUSY Sector" of his model, which is separate from the "Standard Model Sector." There is a "Dark Force" operating in the "Dark Sector," and the photons of the "Dark Sector" are "Dark Photons."[11] The rationale for Sundrum's "Dark Model" is what he called "dark matter sightings"—his interpretation of some recent scientific reports, the "sightings" being unexplained phenomena recorded by satellites, which Sundrum likened to "UFO sightings"!

In addition to all these elements, Sundrum's model also includes a string theory element, called a *brane* (originally derived from the word "membrane") on which the various sectors hang. There is a *hidden sector* living on its own brane and connected to the rest of the model through a "compactified" fifth (or higher-order) dimension of spacetime hypothesized by string theory. Gravity mediates all the interactions among the sectors, because dark matter only makes itself evident through gravitational effects. Sundrum's model also includes a *dark Higgs sector* which gives mass to the *dark photon*. The supersymmetric model and the Standard Model and supergravity mediate between the hidden sector and the dark sector, which are sequestered from each other.[12] Needless to say, this model proved too much even for physicists seasoned in very strange models of the universe. And of course it is too cumbersome and complicated to pass any Einstein test for elegance, but it does give you an idea about the kinds of models of the universe contemplated by some working physicists.

You shouldn't get the impression that supersymmetry was developed to solve the problem of dark matter. Supersymmetry is an elegant extension of the ideas of the Standard Model derived using purely theoretical considerations by talented physicists who subscribed whole-heartedly to the idea of symmetry and tried to extend it further. It just so happens that because supersymmetry requires the existence of many more kinds of particles than are known today, it holds the promise of solving the dark matter problem. It also offers the tantalizing hope of unifying the forces of nature at the very high energy characteristic of an epoch shortly after the Big Bang, and the added bonus of a relatively easy renormalizability of the theory.

Sundrum's elaborate model used not only supersymmetry but also ideas from string theory. Historically, the two theories—string theory and supersymmetry—were related: In fact, supersymmetry was inspired by the first ideas about strings. And an early version of string theory was called superstring theory. Both supergravity, the supersymmetric model that includes gravity, and string theory aim at uniting Einstein's general relativity with quantum field theory.

In the last twenty years, string theory has become one of the most important theories in physics because it is considered elegant and has theoretical unification power, two aspects that have attracted the attention of many bright young physicists. We discuss this theory next.

Chapter 12

Looking for Strings and Hidden Dimensions

My meeting with the discoverer of the strange and beautiful—yet controversial—theory of strings took place in an unexpected location. The Italians sure know how to throw a party. One wouldn't think of Genoa, Italy—a drab, gray port city on the Mediterranean whose days of glory are long gone—as a place where some of the greatest physicists of our time would go to congregate. But during the last ten days in October and the first few days of November 2005, they did exactly that. They all came for a party—the greatest science party ever held—the Festival della Scienza, conceived by civic leaders to attract the world's top brains. This citywide celebration of science, with events taking place simultaneously at many locations throughout the city over a period of more than two weeks, has brought new glory to this otherwise gritty city.

The cab driver picked me up at my hotel at 8:30 p.m., exactly as planned, and we drove up the mountain, high above the commercial area and the port, toward an exclusive neighborhood overlooking the Mediterranean. We turned into a secluded private driveway, and as I got out and approached the gate, an attendant opened it and I was led through a beautiful garden redolent with lemon trees. I could hear voices and laughter from the wide terrace above me. As I came in through the door of the villa, the hostess, the wife of a shipping magnate, welcomed me and showed me to a serving table. "Have some rice with truffles," she said and smiled, and then returned to her other guests.

I stood up from my table, intending to go for a glass of wine, and it was then that I saw him, standing alone on the balcony, his graying mane swept back by the wind from the sea. I walked over to him and said, "You must be Gabriele Veneziano." "Yes," he said as he perked up and smiled. We shook hands. "It was you who invented string theory," I said. He mumbled something in modest protestation, but then nodded. "Well, I didn't quite intend to," he said, laughing, "but I can tell you the story, if you are interested." I assured him that I was.

"I am Italian, of course," said Gabriele Veneziano, leaning on the railing of the balcony, staring at the lights of the city and the ships in the harbor below us. "Now I work at CERN and also at the Collège de France in Paris. I commute between Paris and Geneva." I pondered how tedious this must be, having once made this trip by train myself. I recalled that the French TGV—the supermodern electric rail system that easily exceeds 300 miles an hour—comes to an abrupt end at Dijon, and thereafter the train slows to a crawl up the mountains toward the Swiss border and beyond to Geneva.

"Back in 1968 I was making my first steps in physics," he continued. "I went to Israel, to the Weizmann Institute, where I spent a year as a postdoc. I was sitting in my office one day, thinking about electrons, protons, and neutrons and working with their equations of motion. Then all of a sudden it dawned on me: These particles behaved just like the strings of a violin!" He smiled, noticing my surprise, and continued. "So I plugged some numbers into the equations, and it seemed to work: These particles vibrated much like strings do."[1]

Veneziano is well respected as a scientist, and his positions at CERN and the Collège de France are very prestigious, but he has not yet received the full renown he deserves for proposing string theory. His great idea was soon taken up by others, unleashing a wave of new theories. Edward Witten and Juan Maldacena of the Institute for Advanced Study at Princeton, John Schwarz of Caltech, Michael Green of Cambridge, Leonard Susskind of Stanford, and others took Veneziano's idea and built a comprehensive mathematical-physical theory (or rather, several related theories) around it. This complicated system of equations and derivations has become a major part of modern physics because scientists believed that it held the promise of solving the most important mysteries of the physical world. But unfortunately no results of the theory have been definitively confirmed. The reason for this appears to be that experiments with particles have never reached the energy levels required to prove or disprove the validity of string theory. Once again the hopes of physicists ride on the ongoing research at CERN.

String theory supposes the existence of added dimensions in our universe, and according to the theoretical physicist Jacob Bekenstein of the Hebrew University, the origin of mass does not necessarily lie with the Higgs field. Mass can also appear from a dimension reduction: the idea that a universe that is based on ten or eleven dimensions somehow "breaks down" to a lower-dimensional space of only four evident dimensions. Mass can theoretically emerge from such a process.[2] But we do not know with certainty that the true space-time has ten or eleven dimensions. Also, mass could appear from dimensional reduction without resorting to string theory.

In 1995, Andrew Strominger and Cumrun Vafa, both at Harvard, used string theory to calculate the entropy of a black hole, finding that their results agreed with the formula for black hole entropy derived by Bekenstein and Hawking. The derivation of the entropy of a black hole became the major achievement of string theory. This showed, at least theoretically, that string theory can lead to important results. But string theory is still controversial. "Their only success," the prominent mathematician Sir Roger Penrose of Oxford told me when we talked about string theory in Italy in 2005, ten years after this breakthrough, "is the entropy of a black hole."[3] But many physicists are more sanguine about the power and promise of string theory.

String theory has strong connections to the Yang-Mills gauge idea, the mainstay of modern particle physics; hence it is not divorced from what came before it in theoretical physics. "Gauge symmetry is there because we want to deal with particles with spin and with zero mass—nature seems to like them," Veneziano told me and added that gauge symmetry is the only way to do that. "We don't know why nature chooses spinning massless particles, but it seems to be fond of quantum strings," he said. "Gauge theory and general covariance remove the added degrees of freedom in nature, and so they are a way to get rid of unwanted elements in the theory—a way to remove the bad degrees of freedom," he said.[4] String theory, his creation, mixes well with gauge symmetry.

Some of these ideas originate in what is called the Kaluza-Klein theory. In 1926, the Swedish Jewish physicist Oscar Klein and the German mathematician Theodor Kaluza showed that a unification of Einstein's general theory of relativity with the Scottish mathematician James Clerk Maxwell's theory of electromagnetism (developed in 1864) was possible if one added one more dimension to Einstein's four-dimensional space-time.

What does it mean when a theory has more dimensions than the four we perceive? What are the hidden dimensions of a theory? This is not an easy question to answer. It turns out that statisticians, economists, and sociologists have a better intuition about extra dimensions than do many other people, because they work with multidimensional data. When you have information on many variables, such as in an economic study in which income; wealth; years of employment; educational level; and monthly expenditures on food, entertainment, housing, and transportation are all available, each of these variables can be viewed as a *dimension* of the economic problem under study. A computer analysis is usually undertaken to try to estimate the relations among these variables. In an extension of our geometric intuition about the world, the problem then becomes that of geometry (involving slopes and intercepts): The dimensions are simply the variables under study. The researcher understands the slope of one variable with respect to another simply as a responsiveness parameter: As income rises, expenditure on entertainment rises according to some multiplier of income, for example. The researcher doesn't necessarily look for ways of "imagining" a full space of eight or so dimensions. But in physics, we look for actual physical properties of particles and forces and space, and it is difficult to view dimensions as abstractly as an economist or a statistician might do.

When I was a child, a mathematician gave me an excellent example of what an extra dimension of physical space might be like. He drew a square on a piece of paper and said: "This is a jail, and you are a prisoner inside it. How do you get out?" I thought about this problem for a while, and then he told me, "You just climb out through the *third* dimension!" He pointed to the piece of paper, touched the inside of the square, raised his finger, and dropped it outside the square. "There," he said, "you are outside the walls. You traveled through a third dimension and got out." He then made the next step, explaining to me that it was the same if I wanted to escape from a three-dimensional jail. "Just climb out through a fourth dimension," he said. The idea stuck with me. To me, an extra dimension of space-time is something that I may not see, but which might as well be there. It is an additional dimension that may behave just like the usual dimensions we see in everyday life, and it can get you from one location to another without having to break a wall in three dimensions. It is a means of travel that may exist in some mathematically abstract or hidden setting.

Working with the mathematics of string theory, physicists have developed models in which the equations imply the existence of ten or eleven dimensions. They have thus been led to conclude that space-time itself may have ten or eleven dimensions, rather than four (three of space plus one of time). So where are these dimensions? One explanation is that they are very small, and are "curled up" inside the four dimensions of space-time that we do perceive. If such dimensions actually exist—as contrasted with an economic or statistical or sociological model, where these dimensions are simply a device in a computer analysis—then they may show up if enough energy is applied to space. So there is a small chance that when missing energy is discovered through the work of the LHC, some small and hidden extra dimension of space-time might somehow get "expanded" and appear as a sink into which the excess energy has disappeared without a trace. This is what we mean by "hidden dimensions," and the idea does go back to Kaluza and Klein.

Einstein tried to pursue the Kaluza-Klein model with more than four dimensions for space-time but did not succeed. Nor did anyone else. What was needed was a theory of strings, which would not occur to anyone until forty-two years later.

String theory, according to Veneziano, is automatically a theory of gravity and a gauge theory. Strings like supersymmetry as well, he said. Originally, he proposed string theory in 1968 to explain the behavior of hadrons—pions and protons. There, the strings appeared naturally because of the confinement of the hidden quarks: The internal motion of quarks appeared like the vibrations of unseen strings. Today we know about the existence of quarks, and within this framework, the strings that represent the gluons, the bosons that tightly hold the quarks together, are the fundamental elements of string theory as applied to the behavior of quarks. In this context, the usual fields used in physics become the vibrations of strings.

There have been many articles and a number of popular books about string theory, its promise, and its lack of delivery on that promise. One such book even called the theory "not even wrong." I think that all of these characterizations—both negative and positive—are unfair and misleading. String theory is simply a much more mathematically dependent and mathematically complex theory of the physical world than some other theories. Whether you use string theory, supersymmetry, or the Standard Model often depends on how much mathematics, and what kinds of mathematics, you want to use.

Ultimately, all these modern theories use gauge symmetry, because our best models of the universe depend on these connections between physical reality and the symmetries of the universe through the mathematical theory of groups. But in the case of string theory, there is also a very rich background that is firmly rooted in continuous topological models. In addition to the algebraic structures that convey symmetry, we have manifolds—surfaces that are generalizations of the usual surfaces in our everyday spatial experience (called Euclidean space), and we have certain continuous surface structures called branes (as mentioned, they come from the word "membrane") on which various physical structures depend. The connection between this set of models and the Standard Model and its extensions to supersymmetry is very strong and very evident. And, as we've discussed, there is of course a supersymmetric theory of strings, called superstring theory.

Edward Witten is a physicist who has received the greatest honor in pure mathematics: the prestigious Fields Medal, the mathematician's "Nobel Prize" (since there is no Nobel Prize for mathematics). Witten would often sit at mathematical meetings and ask the speakers irritating questions. In the conference "Perspectives in Mathematics and Physics," held at MIT on May 22, 2009, Sir Michael Atiyah said: "I hope Edward Witten is not in the audience here . . . is he? I would give my ideas to Witten, and in 30 seconds he would give me all the reasons why I was wrong. [Hermann] Weyl used to always ask if [John] von Neumann was in the audience, because he used to do the same to him."[5]

Witten's contributions to string theory are extremely abstract (when I attended a talk he gave at Harvard several years ago, I noticed that most of the people in the audience were pure mathematicians). If his highly mathematical theories, or those of others, turn out to be true, their effect on our understanding of nature would be significant. As we've seen, however, the main problem with these interrelated theories of strings is that they need an immense amount of energy to prove experimentally. There is a chance, however, that some of the predictions of string theory could be proved by the experiments at CERN. The rest will have to await the future.

One aim of string theory is to bring us a final theory of the universe: one that would unite particle theories with general relativity so that all four forces of nature could be modeled within the same framework. Steven Weinberg expressed a hopeful attitude about the unification of the forces of nature and our efforts to decipher the ultimate laws of the universe using new theories. "I don't know if the human race is intelligent enough to decipher the laws of physics," he said, "but I hope that it is."[6] "Will it be in finite time, though?" I asked him. "Maybe it will take thousands of years," he mused. "After all, the Greeks thought up atoms, and it took thousands of years to prove that atoms exist."[7] He scoffed at the notion that perhaps there was some deep philosophical reason, something akin to Kurt Gödel's infamous incompleteness theorems in mathematics, that could prevent us from ever gaining a full knowledge of nature and its forces. But in the meantime, until a "final theory" is complete, we have the very workable Standard Model, which Weinberg and his colleagues have given us.

What we ultimately need, however, is an extension of the Standard Model that would also include general relativity—our theory of gravity. As matter becomes very dense, very massive, and at the same time small in size, so that both gravity and quantum mechanics jointly exert their influence—as happens inside a black hole or under the conditions of the very early universe—a new theory is needed to explain nature. And, so far, we don't have a successful, tested theory that weds general relativity with quantum field theory. What is needed is a unified theory of everything. "What would the final theory be like?" I ventured to ask Steven Weinberg. "Probably something like string theory," he answered. Then he turned off the light and gently closed his office door behind us.[8]

Chapter 13

Will CERN Create a Black Hole?

The cows of Cessy, France, seem to have a natural fear of black holes. Or so I thought when I first saw them. French cows are famously free grazing, which makes their milk taste better than that of their American counterparts, forced to spend their lives in narrow stalls. But these cows of Cessy and the neighboring village of Versonnex, whose range includes the location of Point 5 of the LHC, the site of the CMS detector, were almost leaning against the stand of trees that marked the very edge of their territory—as far away as they could get from the Large Hadron Collider.

But the people of this region, on both sides of the Swiss-French border, are more divided in their opinions about the safety of the large particle collider right under their homes and gardens. Marie Musy, a convivial woman in her fifties from the French city of Dijon, whose husband had worked on the construction of the tunnel for the LEP and the LHC, said to me: "There is no danger at CERN; absolutely not!" Still, many people do have various kinds of fears. Some time ago, CERN held an open day, in which people of the region could visit the laboratory. Thousands more came than the organizers had expected. And the people who did expressed their worries about radiation and about the effects of strong magnetic fields in the area. Fewer people living here seemed to be concerned about the danger of black holes.[1] That fear is apparently more common in areas that are farther away from CERN. In fact, such worries predate the construction of the LHC and originate in the United States.

What *is* a black hole? Generally, a black hole is a dead star of a special kind. It's the remnant of a giant star once it has finished its normal life—burned all its hydrogen, helium, and larger elements in nuclear flames, to the point that it can no longer support its own "weight" through the pressure of the radiation resulting from nuclear reactions and then finally explodes in a giant cosmic blast called a supernova. The dead star then collapses to a tiny size, and because of its immensely concentrated mass, its gravitational pull is so strong that even light cannot escape it. The name "black hole" was invented by the illustrious Princeton physicist John Archibald Wheeler, who reportedly coined this term as a joke.

First proposed more than two centuries ago by the English clergyman John Mitchell, and a few years later independently by the French mathematician Pierre-Simon de Laplace, the modern idea of a black hole came about through the work of Albert Einstein as developed further by Karl Schwarzschild, a German physicist who was the first to solve Einstein's field equations of general relativity.

Einstein's ideas of general relativity were so profound and so revolutionary that they changed physics. For example, his theory could address the question, What happens when gravity becomes very, very strong? This means that the curvature of space-time is very large. In the limit, as a large amount of matter becomes highly condensed together and the force of gravity becomes increasingly more intense, what forms is a space-time *singularity*. That singularity is a black hole—a place where the usual laws of physics break down.

Karl Schwarzschild was a highly gifted German astronomer from a prominent Jewish family in Frankfurt. A patriotic German who signed up for military service in World War I despite being forty-one years old, Schwarzschild received a copy of Einstein's paper on general relativity while he was in the trenches of the Eastern Front.

Schwarzschild set to work. He solved Einstein's equations, and the very first solution had stunning implications. It showed that a black hole was a possibility. The solution of Einstein's gravitational equations was exact, and it specified the geometry of space-time near a massive point. The solution was used to demonstrate that if the gravitational pull of an object was strong enough, then even light could not escape it. Schwarzschild sent his solution to Einstein in Berlin, and the latter wrote him back expressing surprise that his equations could be solved so well. Shortly afterward, Schwarzschild contracted an illness on the front and died in March 1916. The radius of a black hole, the maximum distance from the singularity within which light cannot escape the pull of the black hole, is now called the Schwarzschild radius in honor of its discoverer.

A black hole is something that in theory is "infinitely" dense—its gravitational pull is so strong, in fact, that even light cannot escape it. This is why it appears completely black: It cannot emit light and cannot even reflect any light at all. And it is a "hole" because anything that comes close to it will inevitably fall into it, like Alice into the rabbit hole. In 1971, the first actual black hole was detected in space from the X-rays produced from matter spiraling into it. It is called Cygnus X-1, in the constellation of Cygnus, the Swan. By now we know that there are supermassive black holes in the centers of galaxies, and there are other suspected black holes in our own galaxy.

In 1974, Stephen Hawking made the theoretical discovery of an important property of black holes, now called Hawking radiation (and sometimes Bekenstein-Hawking radiation, reflecting the contributions of Jacob Bekenstein). Because of quantum fluctuations at its surface, a black hole can radiate away its mass, but the process is extremely slow. A black hole that is a dead star can radiate away in 10^{70} (that is, 1 followed by 70 zeros) years.

A black hole also could arise if matter is crushed together to a great density, as is done inside the LHC. The LHC protons can each reach an energy level of 7 TeV—half the total energy that can be generated here for a pair of crashing protons (14 TeV), which is far above any level achieved before in collisions of particles in accelerators. With the LHC we have thus entered an unknown energy zone and are producing collisions never before seen on our planet, except for those from cosmic rays, some of which are believed to have far greater energies.

In early 1999, just before the Brookhaven National Laboratory on Long Island inaugurated its heavy-ion particle collider, called the Relativistic Heavy Ion Collider (RHIC), fears began to spread through certain segments of the population about the possibility that the high speeds and attendant energies produced at RHIC might create tiny black holes. This accelerator, designed to study conditions believed to have been present when the universe was a fraction of a second old, crashes heavy ions of gold or copper traveling at 99.995 percent of the speed of light. This machine's maximum energy level is much lower than that of the LHC (RHIC's maximum achieved energy was reported to have been 500 GeV, which is half a TeV, or 28 times less than the LHC's maximum energy when crashing protons; as we recall, when the LHC smashes lead nuclei, it generates even higher energies).

In March 1999, *Scientific American* published an article about RHIC titled "A Little Big Bang."[2] The report explained that the purpose of RHIC was to study how matter behaved in the very early instants after the Big Bang. Walter L. Wagner of Hawaii saw the *Scientific American* article and remembered that he had read somewhere that Stephen Hawking had theorized that in the first few moments after the Big Bang, miniature black holes were produced by the intense energy of the event. If the Brookhaven accelerator was to re-create conditions that prevailed a short time after the Big Bang, he asked himself, wouldn't it therefore also create minute black holes? And if tiny black holes were produced, could they not increase their size by swallowing up matter and eventually devour our entire planet? Wagner wrote a letter to *Scientific American* protesting the decision to turn on the RHIC, for fear that it might do just that.

The editor of *Scientific American* asked Frank Wilczek, who has since won the Nobel Prize, to answer Wagner's letter. His letter, asserting that the probability of a black hole or other dangerous outcomes was very small, was published along with Wagner's in the July 1999 issue. Frank Wilczek's name thus came to the attention of all who opposed the use of accelerators, something that would haunt him almost a decade later, once the Large Hadron Collider was ready to be turned on.

On July 18, 1999, after the appearance of both letters in *Scientific American*, the London *Sunday Times* published an alarming article titled: "Big Bang Machine Could Destroy Earth."[3] The *Times* article said that the collisions of particles inside RHIC could produce black holes that might swallow the Earth or could produce other dangerous outcomes. A caption asked: "The final experiment?"[4]

Responses to the *Times* article came in from around the world. People everywhere became worried that the Brookhaven Laboratory's new accelerator would end all life on our planet. A reporter on technology issues for the American network ABC News called the accelerator "The Doomsday Machine," accused the laboratory of "playing God," and claimed that a physicist had told him that the construction of RHIC was "the most dangerous event in human history."[5]

Letters were rushed to scientists and lab officials. An eleven-year-old student in a New York private school wrote the Brookhaven director: "I am literally crying as I write this letter," and adult letter writers even compared the completion of the accelerator to the Cuban Missile Crisis in 1962 and the Cold War, saying they were more scared of what RHIC might bring than what they had feared might happen in the worst crises with the Soviet Union. One writer said that the Cuban Missile Crisis, the USSR, and China did not frighten him as much as the "insane" and "genocidal" accelerator on Long Island. And one (unnamed) news organization even called the Brookhaven lab's public relations department to inquire whether there was any truth in the public rumor that a black hole produced by the lab had swallowed the small plane piloted by John F. Kennedy Jr., who died along with his wife, Carolyn, and sister-in-law Lauren Bessette when their Piper crashed into the Atlantic Ocean.[6]

The RHIC accelerator at Brookhaven went on line as planned and more than a decade later does not seem to have created black holes or anything dangerous. But the much higher energy levels attainable by the LHC have caused a resurgence of the old fears, and on March 21, 2008, Wagner and his associate Luis Sancho filed a lawsuit in Hawaii to stop the LHC from starting up. It was dismissed. In Europe, Otto Rössler filed his own unsuccessful suit. But these legal moves received media attention, and people around the world became worried.

Could a black hole be produced this time? And would it swallow the Earth? A few weeks before coming to CERN, I visited Leonard Susskind, one of the world's greatest experts on black holes, in his office at Stanford University. Susskind had recently published his book *The Black Hole War*, about his decades-long argument with Stephen Hawking about the nature of black holes. I asked Susskind what would happen to me if I approached a black hole.

Susskind laughed, and reassured me that the chance of a black hole being produced by the LHC was exceptionally small.[7] Even though I wasn't worried about such an unlikely event, I wanted to talk about this possibility, since I found it theoretically quite interesting. I knew that the probability of a black hole was not zero. In fact, other physicists had told me that the chance that micro black holes would be produced at CERN was measurable—although they hastened to add that such black holes would evaporate rather quickly, within a fraction of a second. But a paper by two Italian scientists published in 2009 indicated that the decay of such micro black holes might not be quite as quick as had previously been thought—they could stick around for a while longer than a millisecond or so. And at the speeds they might have as they emanated from the proton collisions inside the LHC, they could travel some distance before they evaporated—perhaps causing damage.

Susskind, in characteristic fashion, was willing to play with ideas.

He pulled out a yellow pad and drew a large "X" on it. "Do you know what these are?" he asked. "Light cones," I answered. "Yes, that's right," he said. "Let's take one of them."[8]

A light cone represents a section of space-time that is reachable by light (the fastest-moving entity in the universe); it thus represents the accessible part of the universe in a given context. Susskind then proceeded to draw lines on the light cone emanating from the vertex, which represented the space-time singularity in the center of the black hole. "If I am here," he said, pointing to a place on one of the lines, "in a spaceship orbiting the black hole, I would see you freeze on the surface of the black hole, on the event horizon. But you would feel absolutely nothing special as you descended into the depth of the black hole, where of course you'd be crushed to death," he said with a sly smile.[9]

Susskind had discussed such an eventuality in an article he published in *Scientific American* in 1997.[10] His paper, "Black Holes and the Information Paradox," contains the most vivid description I have ever seen of what would happen, step-by-step, to a person falling into a black hole. In it, he tells a story about two professors, Windbag and Goulash, who are mortal enemies; one of them chases the other to the neighborhood of a black hole.

> The horizon [of a black hole] separates space into two regions that we can think of as the interior and exterior of the black hole. . . . The horizon is the place with the (virtual) warning: POINT OF NO RETURN. No particle or signal of any kind can cross it from the inside to the outside. . . . What happens to Goulash, who in a careless moment gets too close to the black hole's horizon? . . . He senses nothing special: no great forces, no jerks or flashing lights. His pulse and breathing rate remain normal. To him the horizon is just like any other place. But Windbag, watching Goulash from a spaceship safely outside

the horizon, sees Goulash acting in a bizarre way. . . . As he falls, the latter shakes his fist at Windbag, but Windbag sees Goulash's motions slow to a halt. Although Goulash falls through the horizon, Windbag never quite sees him get there. In fact, not only does Goulash seem to slow down, but his body looks as if it is being squashed into a thin layer. . . . But Goulash sees nothing unusual until much later, when he reaches the singularity, there to be crushed by ferocious forces.[11]

So if a black hole was somehow produced in the trillions of particle collisions inside a detector of the LHC, and this black hole floated toward me, I would feel nothing. I would have absolutely no warning of my impending doom before I passed the point of no return, and the black hole's horizon encompassed me. Of course, for this to happen, the black hole would have to be very massive, since for a tiny black hole the radius of the horizon is extremely small.

But Susskind noted that under the right conditions a tiny black hole could theoretically be produced; and, he said, if this black hole was charged (carried a net negative or positive electric charge), it could remain at a small size, evaporating only somewhat, then growing in size, and fluctuating around some microscopic size—never to evaporate completely.[12]

"What would it do?" I asked. "Would it drift away?" He answered: "It will be simply a charged particle that will drift toward the center of the earth, attracted by gravity."[13] But it could, of course, move in other directions. Could a micro black hole drift toward me and pierce my body? Or swallow me?

On May 6, 2009, I again went to see Gabriele Veneziano, the brilliant physicist who first proposed string theory in 1968, this time at his office in the prestigious Collège de France in the heart of the Latin Quarter in Paris. We discussed the LHC.

I could sense Veneziano's excitement about what was about to happen to physics when this colossal machine's immense power was turned on. The power of the LHC is so much higher than what we've seen before that anything at all could happen once the switch was thrown.

"There are so many events that could then take place," Veneziano said, "that we had to decide which ones to look for and which ones to ignore. We chose to concentrate on events that result in large-angle scattering." Then he added: "There have been discussions at CERN about looking for evidence of a trigger for a black hole. But I think that the probability that one would actually materialize is very small." Veneziano explained that one model that could allow for the creation of a black hole in the LHC relied on the existence of hidden dimensions of space-time.[14]

If a black hole materializes and evaporates according to Stephen Hawking's formula, Hawking would be very happy. "If a micro black hole was created in the LHC, and if it did not swallow up the world but rather radiated away until it evaporated, as Hawking's work on black holes predicts," Veneziano said, "it would constitute experimental proof of Hawking radiation; this could earn Hawking a Nobel Prize."[15]

At a talk I heard at a high-level particle physics conference held in Boston, "The Large Hadron Collider: Beyond the Standard Model," CERN physicist Fabienne Ledroit-Guillon said that "black hole production is something that has been studied for some years at CERN." She continued: "The black hole mass threshold is around 9.5 TeV." And a graph Ledroit-Guillon showed in her presentation indicated that some production of tiny black holes at the LHC could begin at energy levels between 8 and 9 TeV.[16] For now, in 2010, the LHC is churning out energy below this level—7 TeV. If all goes according to schedule, however, it will reach its maximum level of 14 TeV in 2013.

A report issued by CERN called "Review of the Safety of LHC Collisions"—authored by John Ellis and others of the Theory Division, Physics Department, CERN—used the idea of Hawking radiation to argue for the safety of the LHC. It also claimed that cosmic-ray events preclude the possibility of a black hole in the first place. The report stated:

> The LHC reproduces in the laboratory, under controlled conditions, collisions at center-of-mass energies less than those reached in the atmosphere by some of the cosmic rays that have been bombarding the Earth for billions of years. We recall the rates for the collisions of cosmic rays with the Earth, Sun, neutron stars, white dwarfs, and other astronomical bodies at energies higher than the LHC. The stability of astronomical bodies indicates that such collisions cannot be dangerous. . . . If some microscopic black holes were stable, those produced by cosmic rays would be stopped inside the Earth or other astronomical bodies. . . . Any microscopic black holes produced at the LHC are expected to decay by Hawking radiation before they reach the detector walls.[17]

The report refers to the timescale of the life of the solar system. What is this lifetime? And what, truly, is the real evidence from cosmic rays? Do we know for sure?

I gained a deeper perspective on black holes and the perceived or real danger of one arising at CERN when I visited the black hole expert Jacob Bekenstein at the Hebrew University on April 26, 2009. Bekenstein had locked horns with Stephen Hawking on black holes while he was still a student of John Archibald Wheeler at Princeton in 1971.

Bekenstein had also solved an important mystery about black holes: the fact that they seem to defy the second law of thermodynamics, which says that the entropy of the universe (or any closed system) always increases. Hawking laughed at this idea—he dismissed it as nonsense, and Bekenstein became the butt of jokes. But then Hawking had to eat his words: Bekenstein was proved correct.[18]

I asked Bekenstein, "Suppose black holes will occur, what would they be like? Will they drift to the center of the Earth?" He said that he thought that they would have some momentum as they were created and would move in that direction. "Would they follow a reference frame fixed by the stars [something called Mach's principle]?" I asked. "Yes," he said, "but once they were affected by Earth's gravity, they would go through the Earth, down and beyond, and then back in, oscillating in this way. They could, of course, be extremely small—smaller than an atom—and would thus go through your body with no interaction, and the same would happen as they went through the Earth."[19]

But what if they became bigger? "If a black hole was the size of a molecule," Bekenstein said, "it would weigh as much as a small hill. The impact of such mass alone would kill you."[20] This was a sobering thought. However, I find the argument for safety based on cosmic rays convincing. It would appear that the collisions from cosmic rays, which hit stationary targets—the nuclei of atoms in Earth's upper atmosphere—are different from the head-on collisions of protons inside the LHC. But as Jasper Kirkby, a CERN theoretical physicist, has shown me, there is a way of computing the equivalent impacts of cosmic rays and of the LHC protons on the same scale. And when this calculation is done, it shows that in order to achieve the impact that would be equivalent to that of energetic cosmic rays, a collider four hundred times more powerful than the LHC would be needed.[21] So I think that for the foreseeable future we are quite safe from dangerous black holes. But certainly a small, quickly evaporating one may be possible at our current maximal energy level.

If black holes appeared at CERN and evaporated according to Hawking's formula, not only would Hawking earn a Nobel Prize, but we would have the perfect laboratory for testing the interplay between general relativity and quantum mechanics, since a tiny black hole would entail both gravity and a microstructure. It could lead us experimentally to a theory of everything. So creating a small black hole through the proton collisions of the LHC might not be such a bad thing after all—if only we could keep it under control!

Chapter 14

The LHC and the
Future of Physics

W e are now witnessing a great revolution in the history of science. At the moment when physical theories have almost come to a halt because not many new findings have been made to confirm or contradict the reigning theories in physics, scientists and engineers have completed the construction and testing of the largest machine ever built: a particle accelerator with the capacity to generate energy levels that could only be dreamed of in the past. This machine, the LHC, is now colliding protons at immense energies and it promises to open many new doors to knowledge.

Galileo, Kepler, Newton, and other great minds of the seventeenth and eighteenth centuries gave humanity powerful new ideas about the universe. Each thinker brought society a step closer to an understanding of the nature of the cosmos. The discovery of the rotation of the Earth, the moons of Jupiter, and the force of gravity and all its effects changed the way people viewed the physical world. Further developments had to await the passage of almost two centuries until Maxwell developed the theory of electromagnetism late in the 1800s, followed by the relativity and quantum revolutions of the early twentieth century. Einstein and the quantum pioneers brought us a giant step closer to understanding the riddle of creation.

Then, late in the century, particle physics grew with the Standard

Model, and many findings through the work of particle accelerators pushed this science further. After a lull in discovery, we are now on the threshold of a new age in physics, astrophysics, and cosmology, whose main tool is the Large Hadron Collider.

Smaller particle accelerators have taught us much about particles and forces, and large telescopes such as the Hubble have opened to us new windows on the immensity of the universe. But the LHC offers an experiment in physics and cosmology on a scale that is far grander than anything that was possible until now. The discoveries of the LHC may open possibilities in many directions: the Higgs and the Standard Model, supersymmetry, supergravity, string theory, matter-antimatter, hidden dimensions, or something totally unexpected.

One of the main aims of string theory as practiced today is to bring about a unification of general relativity with quantum field theory. The LHC searches for predictions from string theory, as much as these can be found at the energy levels available through this machine. The collider may find hints of hidden dimensions of space-time, thus providing support for string theory and perhaps an explanation for the weakness of gravity, which may "hide" inside the additional dimensions, thus making it weak in our four-dimensional space-time.

More likely, the LHC will find superpartner particles, hence proving predictions of supersymmetry; such particles may solve part or all of the mystery of the dark matter that permeates the galaxies. Other findings could perhaps help in the solution of the riddle of the dark energy. And the asymmetry in the balance of matter versus antimatter in the universe may be solved through the specialized experiments on B-meson decays at the LHC.

Equally, the experiments with heavy-ion collisions are expected to lead to a wealth of information about the quark-gluon plasma that permeated space shortly after the Big Bang. Most anticipated of all, the Higgs boson may appear in LHC collisions, thus providing the final experimental proof of the validity of the Standard Model. Other experiments may lend support for the Standard Model by explaining better the mechanism of symmetry breaking in the early universe and the action of the Higgs field.

In addition, it is possible that exotic particles—particles that so far exist only in the imagination, or even beyond anything people have imagined—may show up in the LHC. These could also be dark matter candidates. But the true nature of even the most highly anticipated particle of all is still completely unknown: the Higgs boson may not be a single particle. There could be a light Higgs, a heavy Higgs, or a Higgs with completely unexpected properties. This is the beauty of science: We never know what will actually take place until we perform the experiment—in this case, the greatest experiment of all time.

The Techniques of LHC Searches

There are three kinds of emissions that follow the proton collisions in the LHC. One is jets—streams of quarks and gluons. The jets emanate at various angles, depending on the energy and type of reactions produced in the collisions. These angles of jet ejection and the intensities of the jets provide crucial information on the kinds and energies of all other particles produced by the collisions.

In addition to the jets, there are discrete emissions of the free (unconfined) leptons: electrons, muons, taus, and their respective neutrinos. LHC collisions produce a great abundance of neutrinos, but they are not detected—their rate of interaction with other matter is so small that they are generally not seen in LHC collision results.

And finally, there is something called *missing transverse energy*. Energy doesn't have an associated direction, but what is meant here is energy of particles that move in directions that are transverse to that of the colliding beams of protons. This is energy that should be there, by the law of the conservation of energy, but we can't find it—hence it is "missing." The missing energy provides information on hidden dimensions of space, dark matter candidates, and more.

What we have is:

Total mass-energy = jets + leptons + missing transverse energy

Once you sum up all the energy that appears after the collisions—
from the jets and the leptons along the directions that are at right angles
to the beam—that total energy should be zero (pluses and minuses
cancel out) because you've started with zero along these directions. The
difference from the measured amount and zero is the missing transverse
energy. Determining it precisely could lead to a discovery. Typically, the
missing energy lies in the direction of one of the jets.

The data from these events are then subjected to sophisticated statistical analyses, depending on how many jets were produced and how
many leptons appeared—one, two, or more—and of which kinds. Remember that the electron is the lightest lepton other than the neutrinos; then comes the muon; and finally we have the very heavy tau.

These pieces of information provide scientists with indications
about what else may have been produced in the reaction: a Higgs particle, or a supersymmetric partner, or signs of a hidden dimension of
space. From pre-LHC experience with other accelerators, we know
that 50 percent of the interactions in the particle collisions involve the
strong force: lighter quarks and gluons; 20 percent of the products of
collisions include the production of the heavy W and Z bosons and
the very massive top quark. These three particles are believed to be
produced copiously in the LHC. The Higgs may be found in several
different ways: A heavy Higgs will likely decay into a top-antitop pair,
and a light Higgs could decay into two taus. It could also decay into
two Z bosons. As mentioned earlier, in the final step it could produce
four muons (two muon-antimuon pairs).

The search for supersymmetry works as follows. Depending on
whether the supersymmetric particle is neutral or charged, there will
be different patterns of its trajectory inside the detector, which will
help identify it. In particular, a kink in the particle's track, or a disap-
pearing track of a photon, will indicate a chargino. And a nonpoint-

ing trajectory of a photon, one with no set direction, would indicate a neutralino.[1]

———

The aim of physics is to bring us an understanding of the physical universe around us. With the Large Hadron Collider we are trying to solve the ultimate puzzle—the puzzle of creation. And on this quest, physics seems to move in the direction of simplicity and unification. We are attempting to capture the essence of reality in a relatively simple—or rather, the simplest possible—formula. And many physicists believe that this formula, the coveted "theory of everything," will be something that will *unify* all the forces of nature.

"Why look for a single superforce that after the Big Bang supposedly broke down into four separate forces?" I asked Jerome Friedman. "Is it because we have a need to unify things?" "No," he answered, "it's because that is where the forces seem to lead us. And under supersymmetry they actually do meet at one point, if you go far back in time to the Big Bang."[2]

There are physicists who disagree with what is called the reductionist view of physics—the idea that the laws of nature can be captured in relatively concise form in a formula or a set of formulas. Some of them even believe that the laws of physics are not constant.[3] But they are a minority. Most physicists recognize, as Einstein's general theory of relativity has shown, that at least a part of the behavior of nature is described well through a concise and elegant mathematical formulation.

And the beautiful mathematics of symmetry, which forms the fundamental underpinning of the Standard Model as well as supersymmetry, contains something magical: It shows us that nature can indeed be understood through simple and yet highly structured mathematical descriptions. These models are not only elegant, they also have surprisingly high explanatory power.

Nature, it seems, works through unified forces captured by elegant mathematical equations. What we aim to do now is to uncover the *ulti-*

mate formulation of the laws of nature—one whose special cases would be general relativity and quantum field theory, in the same way that Newtonian mechanics is a special case of the general theory of relativity at low speeds and reasonably small masses. And the LHC is our powerful experimental tool in this quest.

Many of the great physicists whose work has been described in this book are still doing advanced research in particle physics, and all of them are hoping that the LHC work will confirm at least some of their theories. Others have retired from much of their life's work and are waiting to see how new discoveries will affect the advances they had made decades ago. Thousands of young physicists involved in LHC work are looking for a big break in their careers, either through a theoretical contribution or an experimental one. All are looking forward with anticipation to the continuing work of the LHC and the results it will produce.

The LHC addresses the most important questions about our existence: What happened in the Big Bang? How did the particles and forces of nature evolve after it? How many dimensions does our universe have? Is there an antiuniverse, or how did the antimatter disappear? How did particles obtain their mass? What is the dark matter that permeates galaxies? What is the dark energy? These are major questions about the creation and structure of the universe we hope to answer.

―――――

In February 2010, the first report on the results of the record-energy collisions that had taken place inside the detectors of the LHC in late 2009 was published. The CMS group, which was first to publish, revealed that in its detector, most particles observed at this high collision energy were mesons—intermediate-size particles that are quark-antiquark pairs. This was valuable preliminary information because it told scientists how to conduct an efficient search for the most interesting events to be observed in the future. It provided a baseline against which to gauge future collision data for the appearance of telltale signs of the

Higgs, supersymmetry, extra dimensions, and other aims of the experiment.

Then, in early March 2010, the LHC was powered up again. I was standing with the scientists inside the CERN Control Center, observing the blue line up on one of the large screens on the wall. It progressed steadily, which meant that the beams of protons were again streaming into the twin tubes of the LHC below us and that they were stable. I glanced around the room and sensed the tension growing as the power was slowly raised. Would the machine keep up with the incremental increase in speed and energy?

Engineers were watching the displays representing the cryogenics of the magnets. The bars were all green, indicating no rise in temperature, but they had to continue monitoring the situation very closely—even a small increase in the temperature of a single magnet would cause it to quench, shattering the entire operation. And CERN could not afford another failure like the debacle of September 2008.

In another section of the control center, two very young scientists, a man and a woman, were watching the data streaming in from the feeder accelerators. These too had to be watched for any potential problems after the three-month slumber of the complete system. If anything went wrong here, the operation of the LHC would have to shut down instantly to avoid damage to the complicated, interconnected system of tubes, electronic components, radio frequency devices, and powerful magnets.

And in the most important area of the center, the northeast section of this large room, Stefano Redaelli was controlling the power of the LHC itself through the computer mouse in his hand. Looking at him serenely leaning over his computer console, I knew I was witnessing history being made by these professionals whom science has entrusted with such an immense responsibility. Behind the young scientists and engineers, the gray-haired Lyn Evans was pacing around, nervously glancing at many of the key displays on the screens above. Would he be able to detect any problem before it turned serious?

A critical stage in the firing up of the machine had passed without incident, and everything worked just as planned: Protons were flying inside the tunnel at record energies never before created by humans. I saw Lyn Evans stretch, turn around, and walk toward the door. He was smiling. The greatest machine ever built was on its way to discovering new physics.

Afterword

At 5:20 a.m. on March 19, 2010, the LHC reached its maximum target energy for the next two years: 3.5 TeV for each beam, or a total of 7 TeV for both beams together. At this energy, the machine is accelerating the protons to 99.9999964 percent of the speed of light.[1] The power will be increased to the maximum total of 14 TeV only after a long period of operation followed by as much as a year of maintenance—now expected to happen around 2013—according to a decision made by CERN management in order to stay on the safe side and avoid the danger of a magnet quenching until the LHC can be overhauled after a year or two of continuous operation.

A total energy of 7 TeV was, of course, another world record, and CERN scientists celebrated the achievement with the traditional champagne. The lab's top management decided that, for the present, the LHC was working well and the chance of an accident was small; it was time to tell the world's press about the planned date for particle collisions at this target energy. CERN notified journalists that the date for starting collisions at a total energy of 7 TeV was March 30, 2010. But the e-mail message to reporters around the world cautioned that getting the beams to collide was a complex operation, which might take several days. "It's like firing needles across the Atlantic, and making them hit each other midway across the ocean," said Steve

Myers, CERN's director of accelerators and technology.[2] More than two hundred journalists from around the globe converged on CERN for this major announced event.

The Big Day: March 30, 2010

During the night of March 29–30, CERN engineers increased the beam energy three times to the target level of 3.5 TeV per beam without causing any particle collisions. Their measurements showed that the machine could well sustain proton beams at this energy for at least a short time. Then the beams were dropped till morning—the protons stopped circulating. Everything was now ready.

At about five a.m. on March 30, the first people started arriving at the CERN Control Center and at the four control rooms of the main LHC experiments. Among them was Manuela Cirilli, who came to the ATLAS control room, where she found other scientists making preparations for the big event. At CMS, Guido Tonelli was coaching his staff, giving last-minute directions as the team was arriving and making ready for the collisions to take place inside their detector.

As the morning progressed, the CERN Control Center was becoming packed to capacity—with scientists doing their work and journalists trying to get a good picture of what was going on. Stefano Redaelli was moving from one computer console to another, adjusting the beam strength and the power output. Many others were performing various tasks. You could feel the tension and excitement in the air. Then, suddenly, the beam was lost. There was a great commotion in the room, and within a few minutes the culprit was identified: the recently installed quench protection system. A slight current variation had tricked the very sensitive devices into thinking that a magnet had quenched, and the system did what it was supposed to do: immediately dump the beam. Everyone waited with anticipation as the process was restarted. The current slowly went up again, and things seemed normal. But later that morning another problem occurred: a glitch in the SPS, the accelerator that directly feeds the LHC with protons. Again the problem was corrected, and the operation resumed. People seemed to be holding their collective breath as the energy was slowly ramped up to 3.5 TeV per beam.

A screen display on the wall now showed two distinct lines: a blue line on top and a red line under it. The top line represented the location of the beam going clockwise around the 16.5-mile tunnel, and the red line the track of the counterclockwise beam of protons. Everyone held their eyes to this screen. The energy level had been stable at 3.5 TeV for a good hour. It was now time to make the beams collide. This was the tensest and most exciting moment in the lives of most of the young scientists here, and for the older ones it was the culmination of twenty years of preparation: from planning to construction to operation of the great machine. They had all been waiting for this one moment, when the LHC would start to collide protons at an energy level never before seen on Earth.

Redaelli and his colleagues now had to get the two beams to collide at the four specific locations of the LHC: the cavities of the four detectors ATLAS, ALICE, CMS, and LHCb. By remotely controlling the magnets at given locations underground below them, the scientists "pushed" the two beams toward each other at every one of the four detectors. This was observed on screen as the blue and red lines slowly began to approach each other: the blue line descending and the red line moving up toward it. This was the trickiest part of the operation: It was, indeed, like firing needles across the Atlantic and hoping to make them collide over the middle of the ocean. People watched with rapt attention. Then, at 12:58 p.m., loud applause erupted in the CMS control room near Cessy: The beams had found each other and collided perfectly head-on. Within a minute the same thing happened at ATLAS, and right after that the two other detectors recorded collisions as well.

At the CERN Control Center, everyone was jubilant. Spontaneous applause erupted, champagne bottles popped open, and people congratulated one another. The scientists and engineers continued to work, holding their champagne glasses. Collimators were now brought down toward the beams, to keep them stable and in place. Protons continued to circulate and collide for hours afterward, producing large amounts of data.

Manuela Cirilli described the situation at ATLAS that day: "It was really different from 2008, when nothing happened. At noon we were all really nervous and tense. There had been a series of minor problems, but nothing really unusual, and nothing really threatening—on a 'normal' day, without the journalists around, we would not even have noticed it. But on this day, we had all the journalists' faces pressed against the glass wall of the ATLAS control room, and we knew that some of them were already typing their articles because they had deadlines at midday to send them to their newspapers. So in this sense it was brilliant: The collisions finally happened just moments before the journalists would have had to send their articles describing the failure of the LHC!"[3]

Interviewed at ATLAS as the collisions were taking place there, Fabiola Gianotti said: "The particles we are now seeing are in huge numbers and with an incredible amount of energy."[4] Peter Jenni, who had directed ATLAS until recently, was moved by the event he had been expecting for so long, saying "This was one of the greatest moments in my professional life. After dreaming about the LHC and its great physics promises for twenty-six years, and then working hard on making ATLAS a reality for seventeen years, it has been an emotional moment for me to see the first high-energy collision events showing up as beautiful displays on the screens in our control room. It gave me much satisfaction to share this moment with many young colleagues and students with great hopes for the physics to come!"[5]

At CMS, the excitement was equally visible. "What is really unexpected," said Guido Tonelli, the head of the CMS group, in describing the first collisions at record energy right as they were happening, "is the frequency of the events. We have been expecting about one unusual event a minute. Instead we are at fifty Hertz—we are getting fifty or more unusual collision results every minute!"[6]

Later, as the celebration continued, Tonelli told me: "This is the moment for which we have been waiting and preparing for many years. We are standing at the threshold of a new, unexplored territory that could contain the answer to some of the major questions of modern physics: Why does the universe have any substance at all? What, in fact, is 95 percent of our universe actually made of? Can the known forces be explained by a single unified force? Answers to these questions may rely on the production and detection in the laboratory of particles that have so far eluded physicists."[7]

Looking forward to the exciting future work he and his team were facing, Tonelli said: "We'll soon start a systematic search for the Higgs boson, as well as particles predicted by new theories such as supersymmetry, which could explain the presence of abundant dark matter in our universe. If they exist, and the LHC will produce them, we are confident that CMS will be able to detect them. We are already starting to study the known particles of the Standard Model in great detail, to perform a precise evaluation of our detector's response, and to measure accurately all possible backgrounds to new physics. We are now starting a new adventure in modern physics. Exciting times are definitely ahead!"[8]

None could conceal their enthusiasm. Lyn Evans said: "Today is the end of a very long period of preparation—over twenty years. But it is also the beginning of a new age in science."[9] Some even compared the new era we are now entering with the LHC to the great revolution that changed physics in the early part of the twentieth century bringing us relativity and quantum mechanics.

The immense energy provided by the LHC, the unprecedented precision craftsmanship that had gone into building it, the wealth of scientific knowledge used in its operation, and the ability, drive, and dedication of the thousands of scientists who run its experiments will undoubtedly bring us new discoveries beyond our wildest imagination.

ATLAS (A Toroidal LHC ApparatuS) is a good example of an LHC detector, and we'll use it to show how a detector works. The seven-story structure is composed of many parts—each designed with a special purpose to identify a certain kind of particle.

The *inner detector* is the inner layer of the ATLAS detector, which surrounds the chamber, the section of the LHC accelerator in which the protons collide. The inner detector maps the tracks of electrons and photons that result from the collisions. Here are thousands of pixels that emit radiation when a particle goes through them, thus allowing their tracks to be recorded. The first detectors inside ATLAS are the pixel detector, the semiconductor tracker, and the transition radiation tracker. These are composed of thousands of electronic modules.

Each module in the pixel detector has a thin outer silicon layer. The semiconductor tracker and the pixel detector work in similar ways. The pixel detector has an outer silicon shell, connected to a lower level of electronics through thousands of little metal spheres. As a charged particle passes through the silicon shell, it liberates electrons, which travel down and induce electric charge in the metal balls. By determining which of the spheres has a charge, which turns into electric current, scientists can determine the track of the charged particle through this inner detector.

The transition radiation tracker, placed outside of the semiconductor tracker and the pixel detector, can distinguish between different kinds of particles. It is made out of thousands of tubes filled with gas. These are all connected with gold wires. When a charged particle passes through the material between two tubes, photons are produced. Electrons and pions (medium-size particles composed of two quarks) have different characteristics as they pass through this part of the detector. A pion ionizes the gas in the tubes and is also accompanied by photons created by this passage. These photons interact with the gas atoms, liberating electrons, which move to the gold wires and are conducted within them. An electron, on the other hand, liberates many more photons, which cause a more intense current to be conducted through the gold wires. This allows the scientists to identify the passing particle, and its track is determined by which of the gold wires have produced the current.

Outside these three inner detectors is the large superconducting magnet whose magnetic field bends the tracks of the particles to identify their charge and momentum. Outside this inner superconducting magnet is the electromagnetic calorimeter, designed to measure the particles' energies, and hence also their masses, as they come out of the inner detectors. The calorimeter is accordion shaped, and it is made of many layers of plates of lead and stainless steel. These are the particle absorbers, and between them is liquid argon, cooled to -185 degrees Celsius (-301 degrees Fahrenheit). Immersed in the liquid argon is a copper grid that acts as an electrode that is used to take measurements of the particles that go through it, to determine their energies.

A high-energy electron emerging from the inner detector will interact with the metal particle absorbers and will result in the creation of many electrons, positrons, and photons. All three types of particles are then measured as they go through the liquid argon, because they ionize the argon atoms. Electrons produced by these ionizations are collected by the copper grid inside it, causing a current. The totality of the measurements of the current and its location—measurements produced by these low-energy secondary particles created by the original high-energy electron—allows scientists to determine the energy of the original electron that resulted from the proton collisions inside the LHC.

The next level of ATLAS as we move outward is the large outer calorimeter, called the hadronic calorimeter. This calorimeter measures the energies of hadrons created in LHC collisions, large particles that are mostly mesons (for example, the pion mentioned earlier). It is a large array of interleaved steel and scintillator sheets. Scintillators are devices that emit light when particles pass through them. When a high-energy hadron passes through a sheet of steel, it interacts with the nuclei of the atoms in the metal, and through these reactions, a shower of secondary particles is produced. When these particles, in turn, go through the scintillating sheets, they produce light. The light is collected in fiber-optic tubes and then converted into electric current. The current signals are fed into the computer and from these measurements the energy of the original entering hadron is determined.

The final outer layer of the ATLAS detector is the muon spec-

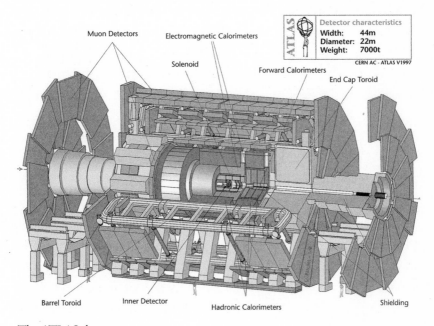

The ATLAS detector

trometer. Muons, which are like electrons but with mass two hundred times larger than that of electrons, usually pass through all the inner layers of ATLAS and reach this outer part of the detector. The muon spectrometer is made of a huge number of chambers, with a total area of several football fields. In each chamber are many tubes filled with gas. When a muon passes through the gas in a tube, it leaves a trail of electrically charged ions and electrons. The electrons drift to the sides and the center of the tube, where they are collected in a metal conductor. Measurements of the induced current allow scientists to determine the paths of muons through the spectrometer.

Appendix B:
Particles, Forces, and
the Standard Model

This appendix summarizes the various particles, the forces and their function, and the Standard Model of particle physics as described in this book.

Particles

Quarks are particles that are too *quirky* to be let out. They make up the protons and the neutrons in the nuclei of matter. They must remain forever confined—inside protons, neutrons, or the short-lived mesons. In the very early universe after the Big Bang, quarks were immersed in the quark-gluon plasma that preceded the formation of normal matter.

Leptons are particles that are free—they can *leap*. These are the electrons and the neutrinos and also the short-lived and heavier muon and tau.

Fermions—named after the Italian American physicist Enrico Fermi—are the "matter particles." They must obey Pauli's exclusion principle: No two of them are allowed to occupy the same quantum state. Fermions have a *fractional spin* (such as ½) in Planck units.

Bosons—named after the Indian physicist Satyendra Nath Bose—are particles that are not constrained by the exclusion principle. They can even form a Bose-Einstein condensate and become one entity with a single wave function, losing their individuality. Bosons have an *integer spin* (0, 1, or 2) in Planck units. Force-carrying particles are bosons.

Hadrons are particles made of quarks held together by the strong force. Hadrons are categorized as mesons or baryons.

Mesons are intermediate-size particles, each made of a quark-antiquark pair.

Baryons are heavy particles made of three quarks each. A baryon with two up quarks and one down quark is a proton, and a baryon made of two down quarks and one up quark is a neutron. Protons and neutrons are also called nucleons when the emphasis is on the fact that they live inside the nucleus of the atom.

Functions

Force-carrying particles are **bosons.** They *mediate the interactions* of the fermions. They are nature's tireless go-betweens, facilitating the behavior of the leptons and quarks (the fermions).

The **photon** (the particle, or wave, of light and similar radiation such as X-rays or radio waves) is the boson that mediates the electromagnetic interactions. The photon carries the action of the electromagnetic force. Two *electrons* interact with each other by exchanging a photon. Light appears when an electron drops from a higher to a lower energy level in its atom.

The **gluon** is a boson that lives inside a proton or a neutron or a meson (or the quark-gluon plasma). It mediates the action of the *strong force* acting inside the proton or neutron or meson. The gluon carries a color charge and exchanges color charge with the quarks.

The W^+, the W^-, and the Z^0 bosons mediate the action of the *weak force,* responsible for some forms of radioactive decay.

The **graviton** mediates the action of the *gravitational force.* But a graviton has never been observed or detected in any way. Because it is part of a theory of gravity, which is not accounted for by the Standard Model, it does not appear in the Standard Model table at the end of this appendix.

Antimatter

All the above particles, as well as others not mentioned here, have antimatter "twins": These are particles that are the same except that their charge (if they have one) is the opposite of their twin's charge.

The **electron** has an antiparticle twin called the **positron**—think *positive* electric charge; the electron's charge is negative. The electron and the positron have the same mass.

The antimatter twin of the proton is the **antiproton**; its electric charge is *negative* (because the proton's charge is positive). The proton and the antiproton have the same mass.

The antiproton is made of three antiquarks: two anti–up quarks and an anti–down quark. The antineutron is made of two anti–down quarks and an anti–up quark.

The electron neutrino has an antimatter twin called the anti-electron neutrino, and so on for all other particles.

When a particle meets its antiparticle, the two annihilate each other.

The Four Forces of Nature

The four forces that govern the behavior of the particles are gravity, electromagnetism, the weak force, and the strong force.

Gravity is the weakest yet long-range force in the universe. It affects everything in the universe, including light (as Einstein has shown).

Electromagnetism is a long-range force, stronger than gravity. Electromagnetism holds the electrons in their orbits in atoms; it also allows atoms to react chemically, producing the molecules of nature and life.

The **weak force** is a very short-range force, acting inside the nucleus of the atom. It is responsible for the radioactive processes called beta decay and similar phenomena.

The **strong force** is also a very short-range force. It acts on the quarks inside a proton, a neutron, or a meson, and its (residual) action is what holds the protons and neutrons together inside the nucleus. This force becomes *stronger* with distance (inside the proton, neutron, or meson), which is why the quarks are confined.

THE STANDARD MODEL OF PARTICLE PHYSICS:

	Fermions			Bosons
Generation:	*I*	*II*	*III*	
Quarks:	up	charm	top	photon
	down	strange	bottom	gluon
Leptons:	electron	muon	tau	Z^0
	electron–	muon–	tau–	W^+
	neutrino	neutrino	neutrino	W^-

A scalar boson (believed to impart mass to massive particles above):

The Higgs

Appendix C:
The Key Physics Principles
Used in This Book

1. Einstein's Law of Mass and Energy

energy = mass times the speed of light squared ($E = mc^2$)

More important in particle accelerator work is a different form of the same law:

mass = energy divided by the squared speed of light ($m = E/c^2$)

This second equation tells us that mass can be created from energy, which is what happens in a particle accelerator: Particles (mass) are smashed with a large amount of energy; then *this energy leads to the appearance of mass*—new particles.

2. Heisenberg's Uncertainty Principle

If you know the momentum of a particle with precision, you don't know its position with precision. If you know time with precision, then there is more uncertainty about energy (or equivalently, mass).

3. Pauli's Exclusion Principle

For fermions (particles with fractional spin, such as electrons and quarks), no two particles may be in the same quantum state. For example, if two electrons are in the same orbital in an atom, then they must have opposite spins.

4. Particle-Wave Duality

Particles can exhibit wave behavior in certain experimental designs; and waves can exhibit particle behavior.

5. Schrödinger's Cat Paradigm, or the Superposition Principle

States of nature can be superposed. Schrödinger's Cat can be in a superposition of two states: dead and alive. In particle physics, this means that a particle can be in a mixture of states of itself.

Notes

All references in this book in the form "AIP" refer to material from the Niels Bohr Library and Archives of the American Institute of Physics, One Physics Ellipse, College Park, Maryland.

Chapter 1

1. One TeV is 1,000 GeV.
2. Author's interview with Guido Tonelli at the CMS control center, March 5, 2010.
3. Actually, for a fast-moving particle, the total energy released is $E = mc^2/\sqrt{(1 - v^2/c^2)}$, where v is the particle's velocity.
4. Rewriting $E = mc^2$, which expresses energy as mass times the square of the speed of light, gives us: $m = E/c^2$, which expresses mass as energy *divided* by the square of the speed of light.
5. "The European Strategy for Particle Physics" (CERN Council Strategy Group, 2006), 15.
6. The LHC's maximum energy is seven times greater than that of the previous record holder, the Tevatron accelerator at Fermilab, near Chicago, which generates just under 2 TeV.
7. The term "The God Particle" comes from the title of a book by the Nobel Laureate Léon Lederman.
8. Author's interview with Jerome Friedman, Brookline, Massachusetts, December 16, 2009.
9. The Big Bang itself was a singularity in space-time, similar in a sense to the space-time singularity at the center of a black hole. Here the laws of physics don't make sense. So if we want to go to "the earliest times," these have to be split seconds after the Big Bang.
10. Frank Wilczek, MIT, "Anticipating a New Golden Age" (lecture given at SUSY 2009—The 17th International Conference on Supersymmetry and the

Unification of Fundamental Interactions, Northeastern University, Boston, June 5–10, 2009).

11. Y. Nambu, *Quarks: Frontiers in Elementary Particle Physics* (Singapore: World Scientific, 1985), 28.

12. Olivier Dessibourg, "Au CERN, la rumeur menace les chercheurs," *Le Temps*, September 10, 2008.

13. Alan Barr, Oxford University, "Sparticle Mass Measurement at the LHC" (lecture given at the International Workshop "Beyond the Standard Model Physics and LHC Signatures," BSM/LHC'09 Conference, Northeastern University, June 4, 2009).

14. Slide of the *Sun* headline from September 10, 2008, in Alan Barr, "Sparticle Mass Measurement at the LHC" (lecture given at the International Workshop "Beyond the Standard Model Physics and LHC Signatures," BSM/LHC'09 Conference, Northeastern University, June 4, 2009).

15. Author's interview with Alan Guth, Cambridge, MA, May 14, 2009.

16. Author's interview with Manuela Cirilli in the CERN cafeteria, March 5, 2010.

17. Author's interview with Stefano Redaelli at the CERN Control Center, March 5, 2010.

18. Author's interview with Manuela Cirilli in the CERN cafeteria, March 5, 2010.

19. "The LHC Enters a New Phase" (news report, January 25, 2010, www.cern.ch).

20. Dennis Overbye, "The Collider, the Particle and a Theory About Fate," *New York Times*, October 12, 2009.

21. John Gunion, "SUSY and the Ideal Higgs Boson" (lecture given at the International Workshop "Beyond the Standard Model Physics and LHC Signatures," BSM/LHC'09 Conference, Northeastern University, June 4, 2009).

22. Author's interview with Paola Tropea at the CERN Control Center, March 5, 2010.

23. Author's interview with Stefano Redaelli at the CERN Control Center, March 5, 2010.

24. http://twitter.com/CERN/statuses/6736202425, a CERN "tweet," December 16, 2009.

Chapter 2

1. Author's interview with Guido Tonelli at Point 5 of the LHC, March 5, 2010.

2. George Smoot and Keay Davidson, *Wrinkles in Time: Witness to the Birth of the Universe* (New York: Morrow, 1993), 22.

3. In fact, during that miraculous year, 1905, Einstein wrote a total of twenty-six scientific articles. See Jürgen Renn, ed., *Albert Einstein—Chief Engineer of the Universe: Einstein's Life and Work in Context* (Berlin: Wiley-VCH, 2005), 92.

4. You can hear the interview, in Einstein's voice, at www.aip.org/history/einstein/voice1.htm.

5. The two numbers are not added, but there is a more complicated formula that combines Einstein's $E = mc^2$ with his principle of special relativity, which says that mass increases with speed, and this formula allows physicists to find the total energy of a moving particle. This formula is obtained by multiplying mc^2 by the Lorentz factor, which is $1/\sqrt{(1 - v^2/c^2)}$.

6. Einstein's letter to Max Born, December 4, 1926, in Max Born, *The Born-Einstein Letters* (New York: Walker, 1971), 91.

7. See, for example, Amir D. Aczel, *Entanglement: The Greatest Mystery in Physics* (New York: Basic Books, 2002).

8. Alan Guth, *The Inflationary Universe: The Quest for a New Theory of Cosmic Origins* (Reading, MA: Addison-Wesley, 1997), 185.

9. Ibid., 286.

10. Ibid., 88.

11. Author's interview with George Smoot by phone to Seoul, South Korea, May 29, 2009.

12. Otto Frisch, *What Little I Remember* (New York: Cambridge University Press, 1979), 57.

13. Isaac Asimov, *Inside the Atom,* 3rd ed. (New York: Abelard-Schuman, 1966), 27.

14. Michael Atiyah, "Geometry and Physics: Past, Present, and Future" (lecture given at "Perspectives in Mathematics and Physics," Singer Conference 2009, in celebration of I. M. Singer's 85th Birthday, Massachusetts Institute of Technology, May 22–24, 2009).

15. A good source on the life of Galois is Mario Livio, *The Equation That Couldn't Be Solved* (New York: Simon & Schuster, 2006).

16. AIP interview with Sheldon Glashow conducted by Timothy Ferris at Harvard University for the television program *Creation of the Universe,* March 27, 1982, 2.

Chapter 3

1. For more on this, see Amir D. Aczel, *Entanglement: The Greatest Mystery in Physics* (New York: Basic Books, 2002).

2. Author's interview with Paolo Petagna, CERN, Switzerland, April 2, 2009.

3. Ibid.

4. Author's interview with Guido Tonelli at Point 5 of the LHC, March 5, 2010.

5. Author's interview with Paolo Petagna, CERN, Switzerland, April 2, 2009.

6. Author's interview with Luis Alvarez-Gaume, CERN, Switzerland, April 2, 2009.

7. Ibid.

8. Ibid.

9. Author's interview with Barton Zwiebach, MIT, March 11, 2009.

10. This is at the full power of 7 TeV for each of the two colliding beams. At lower energies used at the LHC, the speed is still better than 99.998 percent of the speed of light. This information is from the brochure *LHC: The Guide* published by the CERN Communication Group, January 2008.

11. The magnetic field of the superconducting magnets that accelerate the protons around the circuit is much more intense: slightly more than 8 tesla.

12. André David (LIP, Lisbon), "CMS: From Commissioning to First Beams" (lecture given at SUSY 2009—The 17th International Conference on Supersymmetry and the Unification of Fundamental Interactions, Northeastern University, Boston, June 5–10, 2009).

13. Ibid.

14. Author's interview with Paolo Petagna at the CERN cafeteria, March 4, 2010.

15. Paolo Petagna, private communication to the author on December 18, 2009.

16. Author's interview with Fabiola Gianotti, CERN, Switzerland, April 2, 2009.

17. Ibid.

18. Ibid.

19. Ibid.

20. Ibid.

21. Zeeya Merali, "The Large Human Collider," *Nature* 464 (March 25, 2010): 482–84. I am indebted to Manuela Cirilli for bringing this article to my attention.

22. Ibid., 483.

Chapter 4

1. Author's interview with Barton Zwiebach, MIT, March 11, 2009.

2. Author's interview with Paolo Petagna, CERN, Switzerland, April 2, 2009.

3. Peter Jenni, CERN, "LHC Entering Operation: An Overview of the LHC Program" (lecture given at SUSY 2009—The 17th International Confer-

ence on Supersymmetry and the Unification of Fundamental Interactions, Northeastern University, Boston, June 5–10, 2009).

4. Information provided courtesy of the ATLAS Collaboration, CERN.

5. Maria Spiropulu and Steinar Stapnes, "LHC's ATLAS and CMS Detectors," in *Perspectives on LHC Physics,* ed. Gordon Kane and Aaron Pierce (Singapore: World Scientific, 2008), 39.

6. Ibid., 39. The equation is dp/p = $1/BL^2$ where L is the length of path through the detector and B is magnetic field strength.

7. Ibid., 29.

8. Author's interview with Jack Steinberger, CERN, Switzerland, April 2, 2009.

9. CERN Brochure, CERN Public Relations Office, Switzerland.

10. *LHC: The Guide,* CERN Communication Group, January 2008.

Chapter 5

1. AIP, "Interview with Paul Adrien Maurice Dirac by Thomas S. Kuhn and Eugene Paul Wigner, conducted in Eugene Wigner's house, Princeton, NJ, on April 1, 1962," tape 7, side 1, p. 4.

2. These two anecdotes are from Frank Close, *Antimatter* (Oxford: Oxford University Press, 2009), 35. Close remarks that the second story is based on "folklore" and may be apocryphal but is nevertheless in line with the character of both men.

3. AIP, "Interview with Paul Adrien Maurice Dirac," tape 7, side 1, p. 4.

4. Details of the history of Dirac's derivation of his equation are from Steven Weinberg, *The Quantum Theory of Fields,* vol. 1 (New York: Cambridge University Press, 1995), chapter 1.

5. Composite particles that are fermions have half-integer spins, such as ½, ³⁄₂, ⁵⁄₂, etc.

6. Arthur I. Miller, *Deciphering the Cosmic Number: The Strange Friendship of Wolfgang Pauli and Carl Jung* (New York: Norton, 2009), 237.

7. Bing-An Li and Yuefan Deng, "Chen Ning Yang," in *Biographies of Contemporary Chinese Scientists,* ed. Lu Jiaxi, AIP Volume 3, 1992, pp. 183-87.

8. Miller, *Deciphering the Cosmic Number,* 238.

9. Ibid.

10. Brookhaven National Laboratory News, at bnl.gov, February 15, 2010.

11. George Smoot and Keay Davidson, *Wrinkles in Time: Witness to the Birth of the Universe* (New York: Morrow, 1993), 102.

12. Ibid., 103–105.

13. Close, *Antimatter,* 4–8.

14. Ibid., 4.
15. Duncan Steel, "Tunguska at 100," *Nature* 453 (June 25, 2008): 1157–59.
16. Clyde Cowan, C. R. Atluri, and W. F. Libby, "Possible Anti-Matter Content of the Tunguska Meteor of 1908," *Nature* 206 (May 29, 1965): 861–65.
17. Duncan Steel, "Tunguska at 100," *Nature* 453 (June 25, 2008): 1157–59.

Chapter 6

1. Technically, we are dealing with *relativistic* quantum field theory. A quantum field theory in general doesn't have to be relativistic.
2. A. Zee, *Quantum Field Theory in a Nutshell* (Princeton, NJ: Princeton University Press, 2003), 3.
3. AIP oral history interview with Richard Phillips Feynman by Charles Weiner in Richard Feynman's home, Altadena, California, March 4, 1966, session 1:8.
4. Ibid., session 2:71.
5. Author's interview with John Archibald Wheeler, High Island, Maine, June 24, 2001.
6. AIP oral history interview with Richard Phillips Feynman by Charles Weiner, in Richard Feynman's home, Altadena, California, March 5, 1966, session 2:133–34.
7. AIP oral history interview with Richard Phillips Feynman by Charles Weiner in Richard Feynman's home in Altadena, California, June 27, 1966, session 3:5.
8. Ibid., 12.
9. Ibid.
10. Ibid.
11. Ibid., 13.
12. Ibid., 15–16.
13. Ibid., 35–36.
14. Ibid., June 28, 1966, session 4:224–25.
15. Ibid., 225.
16. Mikhail Shifman, *ITEP Lectures on Particle Physics and Field Theory* (Singapore: World Scientific, 1999), 6.
17. Karen Lingel, Tomasz Skwarnicki, and James G. Smith, "Penguin Decays of B Mesons," *Annual Review of Nuclear and Particle Science* 48 (1998): 255.
18. Jacob Bekenstein, "The Fine-Structure Constant: From Eddington's Time to Our Own," in *The Prism of Science: The Israel Colloquium—Studies in History, Philosophy, and Sociology of Science,* vol. 2, ed. Edna Ullmann-Margalit, Boston Studies in the Philosophy of Science 95 (Boston: D. Reidel, 1986), 209.

19. Richard P. Feynman, *QED: The Strange Theory of Light and Matter* (Princeton, NJ: Princeton University Press, 1985), 131.

Chapter 7

1. It is about -1.6×10^{-19} coulombs.
2. Arthur I. Miller, *Deciphering the Cosmic Number: The Strange Friendship of Wolfgang Pauli and Carl Jung* (New York: Norton, 2009), 118.
3. Robert Oerter, *The Theory of Almost Everything: The Standard Model, the Unsung Triumph of Modern Physics* (New York: Penguin, 2006), 145.
4. Luis W. Alvarez et al., "Search for Hidden Chambers in the Pyramids: The Structure of the Second Pyramid of Giza Is Determined by Cosmic-Ray Absorption," *Science* 167 (February 6, 1970): 832–39.
5. Betsy Mason, "Muons Meet the Maya: Physicists Explore Subatomic Particle Strategy for Revealing Archaeological Secrets," *Science News* 172, no. 23 (December 8, 2007): 360–61.
6. Brian Fishbine, "Muon Radiography: Detecting Nuclear Contraband," *Los Alamos Research Quarterly*, Spring 2003, 12–16.
7. Jack Steinberger, "History of the BNL 2nd Neutrino Experiment," unpublished manuscript given to the author by Jack Steinberger, CERN, April 2, 2009.
8. Author's interview with Leon Lederman by phone to Fermilab, March 11, 2009.
9. Author's interview with Martin Perl, SLAC, Stanford, CA, March 24, 2009.
10. Ibid.
11. Frank Close, *Antimatter* (Oxford: Oxford University Press, 2009), 74.
12. Author's interview with Martin Perl, SLAC, Stanford, CA, March 24, 2009.
13. Ibid.
14. Ibid.

Chapter 8

1. This is the Lie group SU(2), the *special unitary group* of order 2, the group of complex-valued 2 × 2 matrices with unit determinant, and with the property that a complex-conjugate adjoint of a matrix multiplied by the matrix itself gives the identity matrix.
2. Frank Wilczek, *The Lightness of Being: Mass, Ether, and the Unification of Forces* (New York: Basic Books, 2008), 43–44.
3. Ibid., 44.
4. These numbers are rough estimates based on a number of recent studies.
5. Author's interview with Mary K. Gaillard by phone to Berkeley, California, June 22, 2009.

6. Ibid.

7. This symmetry is modeled by the Lie group SU(3), the special unitary group of 3 × 3 complex-valued matrices with unit determinant, and with the property that a complex-conjugate adjoint of a matrix multiplied by the matrix itself gives the identity matrix.

8. See Amir D. Aczel, *Entanglement: The Greatest Mystery in Physics* (New York: Basic Books, 2002), 24.

9. Wolfgang Ketterle Nobel Lecture, December 8, 2001, in *Les Prix Nobel 2001*, ed. Tore Frängsmyr (Stockholm: Nobel Foundation, 2002), p. 2 (www. Nobelprize.org).

10. Author's interview with Wolfgang Ketterle, MIT, May 27, 2009.

11. Wolfgang Ketterle, "When Atoms Behave as Waves," Nobel lecture.

12. Author's interview with Wolfgang Ketterle, MIT, May 27, 2009.

13. Author's interview with Jerome I. Friedman, MIT, May 14, 2009.

14. Ibid.

15. Ibid.

16. Author's interview with Jerome Friedman, Brookline, MA, December 16, 2009.

17. Author's interview with Jerome I. Friedman, MIT, May 14, 2009.

18. The word "scalar" means that the field responsible for the particle transforms mathematically as a simple quantity, not a vector. The Higgs is thus called a scalar boson, while the W and Z bosons are vectorial—their fields transform mathematically as vectors.

19. The group is SU(3) × SU(2) × U(1).

Chapter 9

1. According to some theories, the neutrinos may be an exception to this rule, their masses derived from something called a *Majorana* term in their equations, through a process called the seesaw mechanism.

2. These spaces are abstract because the fields and wave functions are complex valued—they have "imaginary" as well as "real" components.

3. Nobel Lecture by Yoichiro Nambu, "Spontaneous Symmetry Breaking in Particle Physics: A Case of Cross Fertilization," delivered by Giovanni Jona-Lasinio of the University of Rome, December 8, 2008, in Aula Magna, Stockholm University (www.Nobelprize.org).

4. Author's interview with Philip W. Anderson, May 25, 2009.

5. Ibid.

6. Ibid.

7. Ibid.

8. Ibid.

9. Ibid.

10. These dates and other information, as well as the full papers, can be found in C. H. Lai, ed., *Gauge Theory of Weak and Electromagnetic Interactions* (Singapore: World Scientific, 1981).

11. Steven Weinberg, "From BCS to the LHC," in *Perspectives on LHC Physics,* ed. Gordon Kane and Aaron Pierce (Singapore: World Scientific, 2008), 139–40.

12. Peter Rodgers, "Peter Higgs: The Man Behind the Boson," *Physics World* 17 (July 10, 2004): 10.

13. Ibid.

14. Ibid., 11.

15. A. Zee, *Quantum Field Theory in a Nutshell* (Princeton, NJ: Princeton University Press, 2003), 236.

16. Rodgers, "Peter Higgs: The Man Behind the Boson," 11.

17. Author's interview with Michael Wick in Boston, June 10, 2009.

18. Author's interview with John Ellis at the CERN cafeteria, March 4, 2010.

19. Steve Connor, science editor, "Higgs v. Hawking: A Battle of the Heavyweights That Has Shaken the World of Theoretical Physics," *Independent,* September 3, 2002.

20. Ibid.

21. Peter Rodgers, "Peter Higgs: The Man Behind the Boson," 11.

22. Connor, "Higgs v. Hawking."

Chapter 10

1. Author's interview with Sheldon L. Glashow, Boston University, May 15, 2009.

2. This is the non-Abelian group SU(2).

3. Sheldon Glashow Autobiography, in *Les Prix Nobel 1979,* ed. Wilhelm Odelberg (Stockholm: Nobel Foundation, 1980), p. 2 (www.Nobelprize.org).

4. Author's interview with Sheldon Glashow, Boston University, May 15, 2009.

5. Ibid.

6. Sheldon L. Glashow, "Partial Symmetries of Weak Interactions," *Nuclear Physics* 22 (1961): 579.

7. Sheldon L. Glashow, *Interactions: A Journey Through the Mind of a Particle Physicist and the Matter of This World* (New York: Warner, 1988).

8. Steven Weinberg Autobiography, in *Les Prix Nobel,* ed. Stig Lundqvist (Singapore: World Scientific, 1992), 1 (www.Nobelprize.org).

9. Author's interview with Steven Weinberg, Austin, Texas, March 5, 2009.

10. Ibid.

11. Ibid.

12. These groups were SU(2) × U(1) × U(1) and SU(2) × SU(2).

13. The unified group is SU(2) × U(1); the broken group was diagonally embedded within the remnants of *both* domains: that of the electromagnetic force and that of the weak force—it was not the original, separate group of electromagnetism alone.

14. Technically, it's a doublet of Higgs fields, each of them complex, so there are two components in each.

15. Author's interview with Barton Zwiebach at MIT, April 2, 2010.

16. Author's interview with Steven Weinberg, Austin, Texas, March 5, 2009.

17. Steven Weinberg, "A Model of Leptons," *Physical Review Letters* 19 (1967): 1264–66.

18. Abdus Salam Biography, in *Les Prix Nobel 1979*, ed. Wilhelm Odelberg (Stockholm: Nobel Foundation, 1980), 1 (www.Nobelprize.org).

19. In 2009, when he was eighty-seven, the French physicist Bernard d'Espagnat won the largest prize ever awarded an individual—the Templeton Prize, valued at $1.42 million.

20. Abdus Salam, "Gauge Unification of Fundamental Forces," Nobel Lecture, Stockholm, December 8, 1979. in *Les Prix Nobel 1979*, ed. Wilhelm Odelberg (Stockholm: Nobel Foundation, 1980), 517–18 (www.Nobelprize.org).

21. Ibid., 518–19.

22. Ibid., 521.

23. Ihsan Aslam, "The History Man," *Daily Times* (Pakistan), August 6, 2004.

24. Malcolm W. Browne, "Abdus Salam Is Dead at 70; Physicist Shared Nobel Prize," *New York Times*, November 23, 1996, Section 1.

25. Aslam, "The History Man."

26. Ibid.

27. Ibid.

28. Gerard 't Hooft, Nobel Prize Autobiography, in *Les Prix Nobel 1999*, ed. Tore Frängsmyr (Stockholm: Nobel Foundation, 2000), 12 (www.Nobelprize.org).

29. Ibid.

30. Ibid.

31. Ibid.

32. Ibid.

33. Author's interview with Gerard 't Hooft by phone to Utrecht, the Netherlands, May 12, 2009.

34. Ibid.

35. Abdus Salam, "Gauge Unification of Fundamental Forces," Nobel Lecture, Stockholm, December 8, 1979, in *Les Prix Nobel 1979*, ed. Wilhelm Odelberg (Stockholm: Nobel Foundation, 1980), 521 (www.Nobelprize.org).

36. Alan Guth, *The Inflationary Universe: The Quest for a New Theory of Cosmic Origins* (Reading, MA: Addison-Wesley, 1997), 136.

Chapter 11

1. There are many references to this; one is a talk by Graham Kribs, "Quirky Dark Matter" (lecture given at the International Workshop "Beyond the Standard Model Physics and LHC Signatures," *BSM/LHC'09 Conference,* Northeastern University, June 4, 2009).

2. Gordon Kane, University of Michigan, "Non-thermal Wino LSP Dark Matter Describes PAMELA Data Well" (lecture given at SUSY 2009—The 17th International Conference on Supersymmetry and the Unification of Fundamental Interactions, Northeastern University, Boston, June 5–10, 2009).

3. This group is SU(5), the special unitary group of order 5.

4. Goran Senjanovic, ICTP, "Proton Decay and Grand Unification" (lecture given at SUSY 2009).

5. Peter Jenni of the ATLAS collaboration at CERN finished his presentation about the promise of the LHC at the SUSY09 conference at Northeastern University in June 2009 with a slide of a drawing of a young woman with the caption: "Maybe we'll find beautiful SUSY?"

6. Julius Wess, "From Symmetry to Supersymmetry" (lecture given at the international conference on supersymmetry, SUSY 2007, Karlsruhe, Germany, July 2007).

7. Osamu Jinnouchi, Tokyo Institute of Technology, on behalf of the ATLAS Collaboration, "Searches for SUSY with the ATLAS Detector" (lecture given at SUSY 2009).

8. Wilfried Buchmüller, DESY Hamburg, "Gravitino Dark Matter" (lecture given at SUSY 2009).

9. Ibid.

10. Fabienne Ledroit-Guillon, CERN, "ATLAS Prospects for Early BSM Searches" (lecture given at the International Workshop "Beyond the Standard Model Physics and LHC Signatures," BSM/LHC'09 Conference, Northeastern University, June 2, 2009).

11. Raman Sundrum, Johns Hopkins University, "Dark Masses and SUSY Breaking" (lecture given at SUSY 2009).

12. Ibid.

Chapter 12

1. Author's interview with Gabriele Veneziano, Genoa, Italy, October 27, 2005.
2. Author's interview with Jacob Bekenstein, Hebrew University, Jerusalem, April 26, 2009.
3. Author's interview with Roger Penrose, Genoa, Italy, November 5, 2005.
4. Author's interview with Gabriele Veneziano, Collège de France, Paris, May 6, 2009.
5. Michael Atiyah, "Geometry and Physics: Past, Present, and Future" (lecture given at "Perspectives in Mathematics and Physics," Singer Conference 2009, in celebration of I. M. Singer's 85th Birthday, Massachusetts Institute of Technology, May 22–24, 2009).
6. Author's interview with Steven Weinberg, Austin, Texas, March 5, 2009.
7. Ibid.
8. Ibid.

Chapter 13

1. Author's interview with Peter Sollander in the CERN Control Center, March 5, 2010.
2. Madhusree Mukerjee, "A Little Big Bang," *Scientific American,* March 1999, 65–70.
3. See details in Robert P. Crease, "Case of the Deadly Strangelets," *Physics World,* July 2000, 19–20.
4. Ibid., 19.
5. Ibid.
6. Ibid.
7. Author's interview with Leonard Susskind, Stanford University, March 24, 2009.
8. Ibid.
9. Ibid.
10. Leonard Susskind, "Black Holes and the Information Paradox," *Scientific American,* April 1997, 18–21.
11. Ibid.
12. Author's interview with Leonard Susskind, Stanford University, March 24, 2009.
13. Ibid.
14. Author's interview with Gabriele Veneziano, Collège de France, Paris, May 6, 2009.
15. Ibid.
16. Fabienne Ledroit-Guillon, CERN, "ATLAS Prospects for Early BSM

Searches" (lecture given at the International Workshop "Beyond the Standard Model Physics and LHC Signatures," BSM/LHC'09 Conference, Northeastern University, June 2, 2009).

17. From the webpage www.lhc.fr. I am indebted to Marie Musy for directing me to it.
18. Author's interview with Jacob Bekenstein, Hebrew University, Jerusalem, April 26, 2009.
19. Ibid.
20. Ibid.
21. Author's interview with Jasper Kirkby in the CERN cafeteria, March 4, 2010.

Chapter 14

1. Osamu Jinnouchi, Tokyo Institute of Technology, on behalf of the ATLAS Collaboration, "Searches for SUSY with the ATLAS Detector" (lecture given at SUSY 2009—The 17th International Conference on Supersymmetry and the Unification of Fundamental Interactions, Northeastern University, Boston, June 5–10, 2009).
2. Author's interview with Jerome Friedman, March 12, 2010.
3. A recent survey article is Adam Frank, "Who Wrote the Book of Physics?" *Discover,* April 2010, 33–37.

Afterword

1. I am indebted to Barton Zwiebach for this calculation of the speed of the protons at 7 TeV.
2. CERN Press Office, e-mail message to registered journalists, March 23, 2010.
3. Manuela Cirilli, personal communication to the author, April 28, 2010.
4. CERN Press Conference, CERN Headquarters, April 30, 2010.
5. Peter Jenni, personal communication to the author, April 7, 2010.
6. CERN Press Conference, CERN Headquarters, March 30, 2010.
7. Guido Tonelli, personal communication to the author, April 2, 2010.
8. Ibid.
9. CERN Press Conference, CERN Headquarters, March 30, 2010.

Bibliography

Note: This listing contains only books; the main sources of information for this book have been professional articles, conference lectures and proceedings, archival material, and personal interviews. References to these sources are found in the Notes section.

Aczel, Amir D. *Entanglement: The Greatest Mystery in Physics*. New York: Basic Books, 2002.

———. *God's Equation: Einstein, Relativity, and the Expanding Universe*. New York: Basic Books, 1999.

Asimov, Isaac. *Inside the Atom*. 3rd ed. New York: Abelard-Schuman, 1966.

Bartusiak, Marcia. *Einstein's Unfinished Symphony*. Washington, DC: Joseph Henry Press, 2000.

Close, Frank. *Antimatter*. Oxford: Oxford University Press, 2009.

———. *The New Cosmic Onion: Quarks and the Nature of the Universe*. New York: Taylor and Francis, 2007.

Davies, Paul. *About Time: Einstein's Unfinished Revolution*. New York: Simon & Schuster, 1995.

Du Sautoy, Marcus. *Symmetry: A Journey into the Patterns of Nature*. New York: HarperCollins, 2008.

Einstein, Albert. *Relativity: The Special and the General Theory*. New York: Crown, 1961.

Evans, Lyndon, ed. *The Large Hadron Collider: A Marvel of Technology*. Lausanne, Switzerland: EPFL Press, 2009.

Fermi, Laura. *Atoms in the Family: My Life with Enrico Fermi.* Chicago: University of Chicago Press, 1954.

Feynman, Richard P. *The Character of Physical Law.* Cambridge, MA: MIT Press, 1967.

———. *QED: The Strange Theory of Light and Matter.* Princeton, NJ: Princeton University Press, 1985.

Frank, Philipp. *Einstein: His Life and Times.* New York: Knopf, 1953.

French, A. P., and Edwin F. Taylor. *An Introduction to Quantum Physics.* New York: Norton, 1978.

Frisch, Otto. *What Little I Remember.* New York: Cambridge University Press, 1979.

Gamow, George. *Thirty Years That Shook Physics.* New York: Doubleday, 1966.

Gell-Mann, Murray, and Yuval Ne'eman, eds. *The Eightfold Way.* New York: W. A. Benjamin, 1964.

Gilmore, Robert. *Lie Groups, Lie Algebras, and Some of Their Applications.* New York: Dover, 2002.

Glashow, Sheldon L., with Ben Bova. *Interactions: A Journey Through the Mind of a Particle Physicist and the Matter of This World.* New York: Warner, 1988.

Greene, Brian. *The Elegant Universe: Superstrings, Hidden Dimensions, and the Quest for the Ultimate Theory.* New York: Norton, 1999.

Guth, Alan. *The Inflationary Universe: The Quest for a New Theory of Cosmic Origins.* Reading, MA: Addison-Wesley, 1997.

Hawking, Stephen. *A Brief History of Time.* New York: Bantam, 1988.

Heilbron, J. L. *The Dilemmas of an Upright Man: Max Planck and the Fortunes of German Science.* Cambridge, MA: Harvard University Press, 2000.

Hermann, Armin. *Werner Heisenberg, 1901–1976.* Bonn: Inter Nationes, 1976.

Hooper, Dan. *Nature's Blueprint: Supersymmetry and the Search for a Unified Theory of Matter and Force.* New York: Smithsonian Books, 2008.

Isaacson, Walter. *Einstein: His Life and Universe.* New York: Simon & Schuster, 2007.

Kaku, Michio. *Hyperspace: A Scientific Odyssey Through Parallel Universes, Time Warps, and the 10th Dimension.* New York: Oxford University Press, 1994.

Kane, Gordon, and Aaron Pierce, eds. *Perspectives on LHC Physics.* Singapore: World Scientific, 2008.

Lai, C. H., ed. *Gauge Theory of Weak and Electromagnetic Interactions.* Singapore: World Scientific, 1981.

Livio, Mario. *The Equation That Couldn't Be Solved.* New York: Simon & Schuster, 2006.

Majid, Shahn, ed. *On Space and Time.* New York: Cambridge University Press, 2008.

Messiah, Albert. *Quantum Mechanics,* vols. 1 and 2. New York: Dover, 1999.

Miller, Arthur I. *Deciphering the Cosmic Number: The Strange Friendship of Wolfgang Pauli and Carl Jung.* New York: Norton, 2009.

Nambu, Yoichiro. *Quarks,* translated by R. Yoshida. Philadelphia: World Scientific, 1985.

Oerter, Robert. *The Theory of Almost Everything: The Standard Model, the Unsung Triumph of Modern Physics.* New York: Penguin, 2006.

Pais, Abraham. *Niels Bohr's Times: In Physics, Philosophy, and Polity.* New York: Oxford University Press, 1991.

Penrose, Roger. *The Large, the Small, and the Human Mind.* New York: Cambridge University Press, 1997.

———. *The Road to Reality: A Complete Guide to the Laws of the Universe.* New York: Knopf, 2005.

Randall, Lisa. *Warped Passages: Unraveling the Mysteries of the Universe's Hidden Dimensions.* New York: HarperCollins, 2005.

Renn, Jürgen, ed. *Albert Einstein: Chief Engineer of the Universe—Einstein's Life and Work in Context.* Berlin: Wiley-VCH, 2005.

Salam, Abdus, and Eugene P. Wigner, eds. *Aspects of Quantum Theory.* New York: Cambridge University Press, 1972.

Schilpp, Paul Arthur, ed. *Albert Einstein: Philosopher-Scientist.* New York: MJF Books, 1949.

Segrè, Emilio. *Enrico Fermi, Physicist.* Chicago: University of Chicago Press, 1970.

Smith, Kenway, Manuela Cirilli, and Heinz Pernegger, eds. *Exploring the Mystery of Matter: The ATLAS Experiment.* Kimber, UK: Papadakis Press, 2008.

Smoot, George, and Keay Davidson. *Wrinkles in Time: Witness to the Birth of the Universe.* New York: Morrow, 1993.

Susskind, Leonard. *The Black Hole War: My Battle with Stephen Hawking to Make the World Safe for Quantum Mechanics.* New York: Little, Brown, 2008.

Trefil, James S. *From Atoms to Quarks: An Introduction to the Strange World of Particle Physics.* New York: Scribner, 1979.

Weinberg, Steven. *Facing Up: Science and Its Cultural Adversaries.* Cambridge, MA: Harvard University Press, 2003.

———. *Gravitation and Cosmology: Principles and Applications of the General Theory of Relativity.* New York: Wiley, 1972.

———. *The Quantum Theory of Fields,* 3 vols. New York: Cambridge University Press, 1995–2000.

Weyl, Hermann. *The Theory of Groups and Quantum Mechanics.* New York: Dover, 1950.

Wheeler, John Archibald, and Wojciech Hubert Zurek, eds. *Quantum Theory and Measurement.* Princeton, NJ: Princeton University Press, 1983.

Wilczek, Frank. *The Lightness of Being: Mass, Ether, and the Unification of Forces.* New York: Basic Books, 2008.

Zee, A. *Quantum Field Theory in a Nutshell.* Princeton, NJ: Princeton University Press, 2003.

Zwiebach, Barton. *A First Course in String Theory,* 2nd ed. New York: Cambridge University Press, 2009.

Photo Credits

Interior

Page 5
Overall view of the LHC experiments: *AC team, CERN*

Page 55
An aerial photograph showing track: *AC team, CERN*

Page 56
CMS detector: *Vittorio Frigo, CERN*

Page 57 and page 231
ATLAS detector: *AC team, CERN*

Page 67
An early "splash" of particles: *CMS collaboration, CERN*

Page 97
A schematic diagram of LHCb: *Rolf Linder*

Page 111
John Ellis drawing a penguin: *Claudia Marcelloni, CERN*

Page 123
Particle tracks: *CERN*

Page 149
Crystals: *USDA*

Page 178
Gargamelle: *Amir D. Aczel*

Insert

The ATLAS detector: *ATLAS collaboration, CERN*

First high-energy collisions inside CMS: *CMS collaboration, CERN*

A muon candidate in ATLAS: *Claudia Marcelloni, CERN*

Particle collisions at 7 TeV in CMS: *CMS collaboration, CERN*

ATLAS 2-jet collision event: *Claudia Marcelloni, CERN*

The CMS detector: *Amir D. Aczel*

Leonard Susskind: *Anne E. Warren*

Stefano Redaelli: *Amir D. Aczel*

Steven Weinberg: *Louise Weinberg*

Gabriele Veneziano: *Amir D. Aczel*

Leon Lederman: *Leon Lederman*

Paolo Petagna: *Amir D. Aczel*

Manuela Cirilli: *Amir D. Aczel*

Guido Tonelli: *CMS collaboration, CERN*

Frank Wilczek: *MIT*

Fabiola Gianotti: *Mike Struik, CERN*

Jerome Friedman: *Jerome Friedman*

Martin Perl: *Jens Zorn*

Philip Anderson: *Eva Zeisky*

Early CMS proton collision: *CMS collaboration*

Aerial photograph with ATLAS and CMS: *Bulletin team, CMS collaboration, CERN*

Index

About the Author

Amir D. Aczel is the author of fourteen nonfiction books, including the international bestseller *Fermat's Last Theorem,* which was nominated for a Los Angeles Times Book Award and has been translated into twenty-two languages. Aczel has appeared on more than thirty television programs, including nationwide appearances on the *CBS Evening News,* CNN, CNBC, and *Nightline,* and on more than 150 radio programs, including NPR's *Weekend Edition* and *Morning Edition.* Aczel is a fellow of the John Simon Guggenheim Memorial Foundation.